✔ KU-525-450

Studies in African History 15

Warfare and Diplomacy in Pre-Colonial West Africa

Studies in African History
General Editor: A. H. M. Kirk-Greene
St Antony's College Oxford

Robert S. Smith

Warfare and Diplomacy in Pre-Colonial West Africa

Methuen & Co Ltd

First published in 1976 by
Methuen & Co Ltd
© 1976 Robert S. Smith
Printed in Great Britain by
Fletcher & Son Ltd, Norwich
Bound by
Richard Clay (The Chaucer Press) Ltd
Bungay, Suffolk

ISBN 0 416 55060 6 hardback
ISBN 0 416 55070 3 paperback

Distributed in the USA by
HARPER & ROW PUBLISHERS, INC.
BARNES & NOBLE IMPORT DIVISION

Contents

Acknowledgements

The writer expresses his gratitude to all those whose generous help encouraged him in writing this book. In particular he thanks the authorities of the University of Lagos, who granted him study leave in the autumn term of 1973, during which most of the reading and some of the writing were done; the Warden and Fellows of St Antony's College, Oxford, within whose hospitable walls that term was spent; the authorities of the University of Aberdeen, who by granting him a teaching and research fellowship in history for 1973–74 enabled him to complete his manuscript and prepare it for publication in the most propitious circumstances; Professor John Hargreaves, whose counsel and support were unfailing, and who read the manuscript, making many valuable suggestions; Dr Roy Bridges, who also read and criticized the manuscript with meticulous care and rendered much help in many ways; Dr Robin Law, who shared his wide knowledge of the subject and also allowed him to use material from two of his unpublished papers; and other friends and colleagues, most of whose names occur among the notes and references. He also acknowledges with thanks the permission accorded him by the joint editors of the *Journal of African History* to reproduce extracts from articles which he published in that journal. Finally, he is greatly indebted to the editor of this series, Mr Anthony Kirk-Greene, for his wise guidance and stimulating suggestions at all stages of the book.

April 1974   R.S.S.

I. Peace and war

Peace and war, those contrary conditions of mankind, are nevertheless alike in one important characteristic, that both are aspects of a society's relations with other societies. They are linked, too, by an intermediate zone in which the tension caused by the interaction of two or more societies is mitigated towards one end of the scale of their relations by peaceful tendencies while towards the other end it is exacerbated by influences hostile to peace. To speak first of peace in this apposition does not imply commitment to the view of Montesquieu that this is man's original condition, nor would the reverse imply support for Hobbes' proposition that his natural state is one of war. Here the intention is to justify the treatment in this book of the relations between West African peoples in pre-colonial times by combining a study of war, that potent – indeed, probably the most potent – vehicle of change, constructive as well as destructive, with a study of diplomacy, which in turn serves purposes both of aggression and of defence and which also may lead to change.

The greater part of the evidence which will be adduced about the relations in peace and war between West Africans, during a period extending over some four or five hundred years to the last decade of the nineteenth century, comes from the histories of those peoples whose levels of political organization entitle their societies to be called 'states'. Arrangements for the exercise of power differed greatly throughout the region, and attempts by social anthropologists to evolve a simple classification or even a 'model' have proved unsatisfactory.[1] But, for the historian's purpose, it remains useful to draw a broad distinction between states where power, and in particular the power to make decisions about peace and war and about negotiations with other states, was concentrated in the hands either of one man or an oligarchy, and those where power was diffused among the members of the community. An example of the first type is the Fon kingdom of Dahomey during most of the eighteenth and nineteenth centuries,[1A] while the Yoruba kingdoms provide examples of the second, and in West Africa the more numerous, type. Reference will also be made, but less often, to the wars and foreign affairs – to use a near synonym for 'international relations' which seems appropriate here – of those people usually described as 'stateless'.

Much of the study of international relations, like the study of governments, has centred around the production of classifications and periodizations. Nicolson, for example, in his short but classic account of western diplomacy, contrasts the 'warrior or heroic' method of conducting these relations with the 'mercantile or shop-keeper' method,[2] while in his reflections on the history of international relations Hinsley sees the year 1763 as constituting in Europe a dividing line between a primitive stage, in which nations seek the physical conquest of other nations, and a post-primitive stage in which the resort to war is more reluctant and more carefully controlled.[3] The relevance of such concepts to the international relations of pre-colonial West Africa may at first seem slight, but in the absence of previous work on the subject they at least provide a point from which interpretation and evaluation of the evidence can proceed.

The study of warfare has similarly evoked numerous generalizations. Much of the attention of writers on warfare (as opposed to military historians, who write about wars) has been directed to the causes of war. As Quincy Wright concludes, 'Wars arise because of the changing relations of numerous variables – technological, psychic, social, and intellectual. There is no single cause of war.'[4] This multiplicity of 'variables', which characterizes most human situations, suggests that the search for generalizations about the causes of war, in pre-colonial West Africa or elsewhere or at any time, has only a limited value and interest. Accordingly, greater stress will be laid in this book on the forms of warfare, especially as these grew out of or were reconciled with the physical features of both the base and the theatre of military operations. The changes wrought by war must also be noted. Its destructive aspect, affecting both material objects and society itself, is plain. On the other hand, many have pointed to the advances in civilization which either owe their origin to or have been quickened by war, stressing in particular the integrative effect of war on society.[5] Wright, writing in 1942, adds a distinction between primitive war which he considers to have been an important element in the development of civilization and wars of historic and modern times which on the whole were and are detrimental to civilization,[6] a stimulating assertion but one which by its nature is incapable of proof or disproof.

An attempt to make a parallel study of international relations in their twin aspects of peace and war must be prefaced by some consideration of law, that concept which together with lying (according to Anatole France) distinguishes man from the animals. Guarding peace at one extreme and mitigating war at the other stands law, that is, the body of rules, whether enacted or customary, which is accepted by a community. So far as individuals within a community are concerned, conduct is regulated by the law of that community. But tension and negotiation between communities, and to some extent war too, fall within the province of 'international law'. Both this and international relations (within whose field the former lies) are concepts evolved in Western Europe where they have been the subjects of much study, but as yet they have been hardly touched upon by students of the indigenous institutions of pre-colonial Africa and of their histories.

The definition of 'law' which is offered above – 'the body of rules, whether enacted or customary, which is accepted by a community' – and which is based upon that given by the *Oxford English Dictionary*, attempts to combine the Austinian or 'command' theory of law with later theories emphasizing its social character. It would thus include the unwritten customary law which prevailed over most of Africa in pre-colonial times and which, Elias has claimed, 'forms part and parcel of law in general'.[7] There was (and is) naturally much diversity in the customary law applied and acknowledged by the many different peoples of the continent. Nevertheless, lawyers have argued that these differences are far from outweighing certain recurring features which distinguish African law. Allott[8] has analysed the most important of these features: the traditional and popular character of the law; basic similarities in judicial procedure, whether in the courts of chiefs or in the arbitral tribunals of villages, clans or households; the role of the supernatural; patterns of government which almost always rest ultimately on the consent of the governed; and the role of the community in the interpretation and enforcement of law. Thus, he concludes, African law is 'a unity in diversity'. Elias, in a contribution to a symposium on 'Sovereignty within the Law',[9] develops this theme, arguing that 'in large areas of Africa there ... emerged broadly similar rules of customary law, which makes it possible to speak of the existence of a universal body of principles of African customary law that is not essentially dissimilar to the broad principles of European law.' He continues that this customary law 'shares with customary international law the characteristic that its validity does not depend upon any theory of sovereignty'.[10]

The existence of a widespread customary law was noticed by early European visitors to Africa, in particular by the Portuguese. Traders enjoyed, for example, the protection of law for the unwritten contracts, based on oaths and 'medicine', which they made with the local inhabitants. Conversely, they resented the application of a West African 'droit d'Aubaine' under which the property of a dead colleague or their ships gone aground along the coast were forfeit to the local ruler.[11]

Consideration of customary law by legal historians has usually formed a part of their consideration of 'primitive' law. By no means all early law falling under this denomination is unwritten, just as by no means all custom is a part of customary law. Diamond's account of what constitutes primitive law is based on the type of economy practised by the society for which the law exists, and he describes it as a law which grows and changes of itself without the aid of or need for legislation.[12] He characterizes most West African states of pre-colonial times, including the Islamized countries, as 'peoples of the late codes' who were at a stage of development comparable to that of the Babylonians in the time of Hammurabi (*c.* 1750 B.C.), the Romans of the era of the Twelve Tables (the mid-fifth to the mid-third centuries B.C.), and England in the twelfth and early thirteenth centuries A.D.[13] These far-reaching comparisons are not in the present case particularly helpful. More to the point is the evidence that this customary law, whose broad similarities transcended tribal boundaries, provided a bond between the different states and peoples of West Africa in pre-colonial times and a form of international law by which their relations with each other could be, and normally were, regulated. The diversity which existed between the different political systems of West Africa serves too to emphasize the importance of customary law in bridging the gaps between them. Moreover, though the evidence for these relations derives mainly from the more centralized states, it must be recollected that the small-scale or 'stateless' societies of West Africa were often linked by associations, such as the Poro of Sierra Leone and Liberia, which observed an elaborate common code of rules and performed political, including diplomatic, functions on behalf of the whole group.[14]

It would be unprofitable to discuss whether customary international law first arose in conditions of peace or of war since its origins varied from place to place and from one period to another, some primitive people being peace-loving, others not.[15] Yet war has rarely entailed a complete repudiation of the restraints of peace and much of modern international law is in fact based on rules of war. Among most West African peoples the regulation of peaceful contact with outsiders was accomplished by some regulation also of their warfare, often constituting a mitigation of that warfare. Perhaps the outstanding example of this is in the formal declaration of war which was widely practised, to the detriment of surprise in attack.[16] Though examples will be cited of countries overrun by their enemies and of their dynasties being ejected, 'unconditional surrender' seems not to have been a usual object of West African wars, nor does the long-term occupation of foreign territory; outright victory in war was often followed by the establishment of a tributary relationship which left those of the indigenous cultivators who survived in possession of their farms.[17]

As fundamental to an understanding of the relations between the peoples of pre-colonial West Africa as the laws they observed is the physical setting within which they lived, and this also must now be briefly described. The northern part of the region consists of a broad belt of grassland running east and west. Known to travellers and historians as the Central and West Sudan, this is further classified by geographers as belonging to three separate types of savannah: furthest north, where the rainfall is scanty, is the Sahel, an area characterized by its stunted thorn trees fairly widely spaced and which merges on the north with the Sahara desert; below this comes the Sudan proper, similar in vegetation but with a rather higher rainfall, and further south again is the Guinea savannah, a land of deciduous woods and long grasses, much broken up by rocks and hills. The southern part of West Africa is largely covered by tropical rain forest which merges into the freshwater and then the brackish mangrove swamps or the sand dunes of the coast. An important exception to this pattern is the 'Dahomey gap' or 'Accra corridor' in which grassland extends south to a strip of coast running roughly from the region of Accra on the Gold Coast to Badagry. The grasslands provide the most favourable conditions for cultivation and habitation, while the swamps and the Sahel are lightly populated as are also the densest parts of the rain forest. It seems likely that in early times the forest and also the hilltops of the Guinea savannah provided a refuge for peoples driven away from their homes by their more aggressive neighbours in the grasslands.

The warfare of the savannah differed from that of the south in the comparative ease and speed with which troops could be moved across country, the open formations which could be adopted both on the march and in battle, and the scope afforded to the use of long-range weapons. The introduction of the horse sometime before *c*. A.D. 1000[18] greatly increased men's mobility, though the prevalence of the tsetse fly prevented the rearing of horses in the Guinea savannah, as it also did in most parts of the forest (a point to which further reference is made in Chapter V). The main weapons of savannah armies were the spear, the javelin and the bow and arrow.[19] By contrast, movement in the forest was along narrow, twisting and impeded pathways, requiring the adoption of 'Indian file' and giving little scope for mobility or tactical deployment. Close fighting and reliance on the ambush were dictated in these conditions, while the sword and the club were the important weapons, though the bow, including the crossbow, was also used. Finally, though there is no trace of any fighting at sea by those few of the coastal peoples who were prepared to venture in their boats through the Atlantic surf, a form of naval warfare existed on the rivers, creeks and lagoons of the south where the war canoes operated not merely as transports but also as fighting units which engaged in battles on water in which, before the introduction of firearms, spears and clubs seem to have been the principal weapons.

Environmental influences can be perceived in the peaceful relations between the West Africans as well as in their warfare. In particular, the greater ease of movement in the grasslands meant less isolation for those living there, and thus political and economic contacts developed over long distances. By contrast, the forest dwellers were to some extent shielded and inhibited by their environment from contact with their neighbours, and this comparative isolation made for wariness and suspicion of outsiders. Yet from early times there was commerce between forest and savannah; kola nuts, for example, are said to have been introduced into Hausaland from Guinea about the middle of the fifteenth century.[20]

Both the peacetime relations and the warfare of the West Africans were affected in pre-colonial times by two major external influences. The earlier of these was Islam, coming across the Sahara from North and North-East Africa from the eleventh century onwards, or even before, and affecting mainly the savannah region. The second was that of Western Europe, whose officials, traders and missionaries began to arrive from the sea in the late fifteenth century and whose impact was confined mainly to the coast. In four important ways these two influences followed similar lines. In the first place they brought a knowledge of reading and writing, knowledge which for long was mastered only by a small élite of literates who thus tended to acquire an importance in their societies which was not based on traditional qualifications for office. Secondly, West Africa was brought into touch with the outer world: with the Berbers, Arabs and Egyptians of North and North-East Africa and with the Ottomans on the one hand, and with Portugal, Spain, France, Holland and Great Britain, the great seafaring and trading nations of West Europe, on the other, and early religious teaching and commercial dealing were followed and fostered by formal and official relations. Thirdly, new ideas were received along with literacy. This somewhat imponderable factor is best illustrated by the manuals of religious practice, statecraft and warfare which were circulated, copied and even written in the Islamic north of the region. European influence can probably be most clearly seen in the spread of new trading techniques, especially the giving and receiving of credit, the keeping of accounts, and the stimulation of an ever more sophisticated market. Fourthly, the conduct of war was affected in many ways, above all by the introduction of firearms. This last was a very gradual process which began in the sixteenth and seventeenth centuries but in many places did not attain importance until the eighteenth and nineteenth centuries. Almost from the beginning the trade in these weapons was dominated by the Europeans, though a few guns reached the region across the Sahara. Underlying all these developments and interacting with them was the export of slaves from West Africa, both to the north across the Sahara and, on a greater scale, to the plantations of the New World.

Thus the foreign relations of West Africa were not confined within the region itself. Throughout the period from about A.D. 1400, the point at which a reasonable degree of evidence begins to be available to the historian, to the beginning of the colonial period in the late nineteenth century, there was almost continuous contact between West Africa and the distant countries of North Africa, the Levant and Western Europe. This contact was mainly peaceful, the only violent interruption on a grand scale being the Moroccan invasion and overthrow of Songhay in 1591. It must nevertheless be remembered that much of the unrest which characterizes the history of pre-colonial West Africa was stimulated by the export trade in firearms, behind which was the foreign demand for slave labour. Finally, external influences must not be exaggerated. Underlying both the Islamic way of life and European innovations, both of which tended often to be mere veneers on African society, was an essentially West African culture, or, more accurately, congeries of cultures. Succeeding chapters of this book, while taking account of and attempting to assess outside contributions, will be directed primarily to examining the indigenous forms taken by the foreign relations and the warfare of West African states during the five centuries which preceded the partition of the area and the establishment of colonies there.

II. Peaceful relations

International relations constitute an inescapable feature of any government's duties. (Lenin's startled question in 1917 – 'What, are we going to have foreign relations?'[1] – was an almost incredible naïvety.) Diplomacy, which is the fundamental means by which foreign relations are conducted and a foreign policy implemented, far from being an invention of capitalism or of the modern national state, or of classical antiquity either, is found in some of the most primitive communities[2] and seems to have been evolved independently by peoples in all parts of the world. The basic object of diplomacy is to enable men to live with their neighbours, a feat which requires a measure of accommodation to the interests of others. The more detailed ends of diplomacy are equally universal – or so nearly so as to raise no substantial question for the historian. Above all they are the questions of peace and war, and then such matters as the conclusion of treaties, the making, maintenance and breaking of alliances, the establishment of boundaries, the development and protection of trade, and the payment of tribute. The means by which these are pursued need to be adjusted to changing circumstances, but the employment of accredited agents – diplomatists – to represent and to negotiate on behalf of a state or society seems to be a nearly constant practice. From this it usually follows that diplomatic forms become established among neighbours and immunities recognized. If common patterns were to be observed in this area of political life among the peoples of West Africa, this would provide a further argument for the tendency towards unity of the customary law over a large part of Africa and for its claim to constitute international law.

The evidence for the conduct of international relations in pre-colonial West Africa before the nineteenth century is unhappily meagre. Yet it is clear from the reports of early travellers that long before then West African states were in the habit of sending their representatives on diplomatic missions to each other. These officials are usually described as 'ambassadors', though other words used are 'messenger', 'linguist' (especially on the Gold Coast), and occasionally, perhaps with a recollection of the evolution of Western diplomacy, 'herald'. The earliest reference to diplomatic relations in West Africa seems to be al-Saghir's account of the sending of an ambassador by a ninth-century Imam of Tahert (in North Africa) to an unnamed Sudanic state, possibly Gao. Then follows al-Bakri's description of the mosque in the royal capital of eleventh-century Ghana which was set aside 'for the use of Muslims who visit the king on missions', and there are indications that political relations were maintained in the late eleventh or early twelfth century between Ghana and the Almoravids in Morocco. In the sixteenth century there is Leo Africanus's mention of the reception at the court of Timbuktu of 'ambassadors from other princes'.[3]

Most of the early references (though not necessarily that by Leo Africanus) are to relations between West African states and states outside the region – either with the Muslims of North Africa in the case of the West and Central Sudan, or with Western Europe in the case of the coastal states. These more distant contacts will be considered in a later part of this chapter, the intention here being first to focus attention on the indigenous forms of diplomacy and on relations between states within West Africa itself. Indications of diplomatic activity under this heading occur in the correspondence of some of the first Christian missionaries to visit the region. In a letter of 1539 to their king, John III, three Portuguese friars at Benin wrote that the Oba (king) there had the 'habit of ill-treating and imprisoning all ambassadors of kings who send messages to him', treatment which he had accorded to the envoys of the two small coastal states of Ardra and Labedde.[4] Another ecclesiastical report, written by the Prefect of a group of Italian missionaries on the island of St Thomas to the Sacred Congregation at Rome, states that in 1691 relations between Benin and the Itsekiri kingdom of Warri were so strained that 'they (were) not exchanging ambassadors',[5] which suggests that in the usual course there was regular diplomatic liaison between the two kingdoms. References to the activities of ambassadors occur throughout the works of Dapper, Bosman, Snelgrave, Norris, and Dalzel, writing in the seventeenth and eighteenth centuries. In the last three, the ambassadors of Oyo, the most powerful of the Yoruba kingdoms, and of Dahomey, figure prominently. In one account it is related that before embarking on his invasion of the kingdom of Whydah in 1727 – the beginning of his drive to extend his rule to the coast – Agaja of Dahomey sent an ambassador to Whydah with a request for 'an open Traffick to the sea side'.[6] Elsewhere Snelgrave relates how in 1730 Agaja 'sent Embassadors with large Presents' of coral 'together with one of his handsomest Daughters' to the Alafin (king) of Oyo. They succeeded in obtaining for their master 'an advantagious Peace', which the Alafin confirmed by sending one of his daughters in return as a wife for Agaja.[7]

The states on and in the hinterland of the Gold Coast were also engaging in considerable diplomatic activity in the eighteenth century, as appears from the reports of contemporary European observers. In 1714, for example, the Ashanti sent a delegation to compose differences between John Kabes, the merchant prince and virtual ruler of Komenda, and the Twifo.[8] Some years later the ambassadors of the ruler of Wassa succeeded in bringing about a league of the coastal states between Cape Appolonia and the Volta in order to prevent the supply of guns to the Ashanti.[9] Another alliance against the Ashanti was engineered by Fante diplomacy on the death of the Asantehene Opoku Ware (*c*. 1750).[10] In 1777 a new Asantehene, Osei Kwame, sent ambassadors to the King of Wassa asking him to ascertain whether the Fante were willing to accept the presents usually made on a king's death – an interesting example of negotiation through a third party.[11] For the nineteenth century, evidence is still more abundant, though the influence of and interaction with Europeans is more intrusive. A late example of indigenous diplomacy from the Yoruba country is the part played by embassies sent by the Ekiti and Ijesha kings to other monarchs in the formation of the anti-Ibadan coalition of 1878 known as the Ekiti-parapo.[12]

The majority of the examples above are of embassies which were concerned with political questions, and they illustrate not merely the maintenance of foreign relations among West African states but also the evolution and implementation of foreign policies. Almost equally important, and equally illustrative of longer-term aims, was the development of commercial relations. Indeed, in West Africa as elsewhere trade and politics played as great a part in foreign relations as in internal affairs. Nothing so clearly indicates the interdependence of the peoples of West Africa as the long-distance trade which linked the coastal and forest states with the savannah lands far to the north. The bulk of such trade seems to have been under the control of the more centralized states,[13] whose governments were of a nature to distinguish between foreign and domestic policies. The trade in kola nuts, a product of the Guinea forest much in demand by the Muslims of the West Sudan, illustrates the connection between commerce and diplomacy.[14] An important part of the trade passed from Ashanti through Gonja, Dagomba and Borgu to Hausaland and was under the supervision of resident representatives of the importers. Lovejoy writes of these Hausa and Hausa-speaking officials – who after the 1830s were restricted by the Kumasi government to Salaga and markets north of the Volta – that they 'served as diplomatic and commercial representatives of the trade network to local states',[15] and indeed their functions seem to have been comparable with those of the early European consuls – for example, those of Venice in England and the Levant.[16]

Commercial relations played a part not only in the development of *ad hoc* diplomacy and in the expansion of foreign relations into a deliberate and long-term foreign policy, but also in the tentative steps which were taken in pre-colonial West Africa towards permanent or continuous diplomacy. In Western Europe the system of maintaining resident representatives at foreign courts, which had its origin in fifteenth century Italy, was the most prominent feature of the new forms of diplomacy evolved by the modern territorial states. Together with organized chanceries acting as embryonic foreign offices, this made possible a continuous and informed foreign policy which was designed, as Mattingly puts it, 'to make diplomacy do the work of arms, to make the foxes masters of the lions', and was the counterpart of the standing armies of the nation states.[17]

The practice of maintaining resident representatives abroad was not unknown to the indigenous diplomacy of West Africa.[18] According to Leo Africanus, the Askia (king) of Songhay in the early sixteenth century 'had some of his courtiers perpetually residing at Kano' for the receipt of the tribute due to him from that kingdom.[19] At the end of the seventeenth century the king of Denkyira appointed an official named Ampim as his resident trade representative on the coast. On his death at Cape Coast in 1698, the English company sent consolatory gifts to the king.[20] In the eighteenth century the Ashanti maintained residents for the collection of tribute in Dagomba and other neighbouring states, and also sent trade representatives to Accra and Akuapem. Their representative at Cape Coast in the early ninteenth century, referred to by Dupuis as their 'Captain general', was accorded considerable respect as 'the organ of his government'. When the holder of the post, 'a young athletic man', died in 1820 (allegations of poison naturally being made), a nephew of the Asantehene was appointed in his place and made his entry into Cape Coast 'with a degree of military splendour unknown there since the conquest of the Fantee by the King' and accompanied by a suite of some 1,200 persons.[21] Similarly, Dagomba and Gonja maintained representatives in Kumasi, the Ashanti capital.[22] There are indications that the Alafin of Oyo appointed ambassadors to pay extended visits to, and possibly reside in, Dahomey in the latter part of the eighteenth century in order to collect the tribute due to him under his treaties and to report any Dahomean military successes so that he might demand a share of the spoil.[23] The Oyo government also stationed agents (*Ilari*) in Egbaland while it was tributary to them, a relationship which was probably broken towards the end of the eighteenth century. Similarly, the Oba of Benin placed agents in such peripheral parts of his territory as the Yoruba town of Akure, where they were known as *Balekale* or *Abilekale*. The Oyo system of residents was extended by the successor state of Ibadan, their representatives in the outlying parts of their empire being known as *Ajele*.[24]

Many of these examples, it will be noted, apply to relations between a dominant power and a tributary. Nevertheless, they are all approximations to continuous diplomacy. They seem, moreover, to have arisen spontaneously out of the indigenous political and economic situation. Another form of continuous diplomacy particularly prominent in West Africa and almost certainly of indigenous origin was the representation of a state or community abroad by chiefs or officials of the foreign state. The system has some resemblance to that of the *proxenoi* in classical Greece or to the use in modern times of honorary consuls. Apparently from early times Hausa towns had among their chiefs a *sarkin Turawa*, 'head of the white men' or Arab traders,[25] an official who was probably originally himself an immigrant. In the late nineteenth century the French agent, Mizon, was received at Yola, capital of Adamawa, by this chief.[26] Similarly, a *sarkin Nupe* looked after Nupe affairs in Hausa towns, and doubtless other chieftancies of this kind existed according to the needs of time and place. At Abuja, whose governmental system was based on that of the Hausa or Habe kingdom of Zaria before the Fulani conquest, an important royal official was known as the *bakon Bornu*, or 'envoy to Bornu'. Barth mentions that there was 'a sort of consul of Bornu' at the court of the Sultan of Bagirmi in 1852 and a consul of Bagirmi, the Mestrema, at Kukawa – the first being a Bagirmi and the second a Bornu citizen.[27] In Dagomba a chief of the Ya Na's court, the Kunkuma, protected the interests of the Ashanti.[28] Another development which seems to have evolved from contact with European traders, was the custom at Badagry, Whydah and some of the Dahomean towns of designating chiefs to guard the interests of European residents and visitors, and often also to provide for their accommodation. Thus, for example, there are references to the 'French', 'Portuguese', 'Dutch' and 'English' chiefs at Badagry, and on arrival there in 1850 Bowen, the American missionary, was asked for a present by the 'American chief'.[29] At Whydah in 1863 Burton encountered the Hun-to, 'nominal head of English town' (and great-grandson of an English corporal), at Kana he was entertained by the Buko-no or 'English landlord', and at Abomey, the capital, he transacted business through the 'English mother', a female official of the court.[30] Similarly,

European traders in the Mende and Temne countries of Sierra Leone were protected by local chiefs or their nominees on the lines of the landlord and stranger-tenant relationship established there with immigrant African farmers.[31]

Little direct evidence can be found as to the location of responsibility for the conduct of foreign affairs among the West Africans. It can be assumed, however, that this must have varied in accordance with the exceedingly diverse forms of government which prevailed in pre-colonial times. Some trace can be detected of the theory commonly found in Europe that such matters were primarily the concern of the executive arm of government. In a monarchy the king usually exercised at least nominal authority over foreign affairs.[32] An Ashanti declaration of war had to be pronounced by the Asantehene seated upon the *hwedom*, a throne reserved for this purpose.[33] According to Johnson, writing of Oyo, 'the control of foreign affairs (was) vested in the King',[34] and any consultation with the council known as the *Oyo Mesi* only implied that the Alafin was uninterested in the matter under discussion. On the Gold Coast, a palaver about foreign affairs was proverbially described as 'a bringing together of stools'.[35] Yet only in a military despotism, such as Dahomey, can it have been possible for a ruler to embark on major moves in foreign policy without the advice of at least his greater chiefs. Even here the royal power to act alone may have been exaggerated, and by the mid-nineteenth century it seems that, as Burton reported, the Dahomean chiefs had largely succeeded in isolating the king from foreign influences.[36] In metropolitan Oyo military power was exercised by the Oyo Mesi, whose leading member, the *Bashorun*, commanded the army of the capital, so that it is difficult to accept Johnson's view of the royal prerogative in foreign affairs. In Ashanti, after what Wilks has called the 'Kwadwoan revolution' in central government during the eighteenth century, the political influence of the aristocracy is said to have declined and that of the Asantehene to have become preponderant. Nevertheless, Bowdich, who was well-placed to assess the issue, considered that 'The constitution requires or admits an interference of the Aristocracy in all foreign politics, extending even to a veto on the King's decision; but they watch rather than share the domestic jurisdiction.'[37]

The status of those chosen to carry out diplomatic duties in West African states varied, but in every case they must have been among those close to the rulers of the country, often members of the royal household. Sometimes they were the great men of the land; even princes from the royal family might be sent on a mission, as in the Congolese embassy to Rome in 1514[38] or the Ashanti representative at Cape Coast referred to above. On one occasion at the end of the seventeenth century the kings of Denkyira and Ashanti sent some of their wives as ambassadors to each other, a courtesy which was said to have provided a cause or pretext for war as the Asantehene, the great Osei Tutu, claimed that his Denkyiran brother had misbehaved towards one of the envoys.[39] Among the Ibo, priests were appointed as ambassadors in negotiations to end the small-scale inter-communal wars.[40] In other cases slaves might be used. This was apparently so in the embassy which the Ashanti proposed to send to England in 1821.[41] Nevertheless, it is unlikely that Dupuis had been misled in his belief that these ambassadors were men of rank and influence since the more important palace slaves at Kumasi, as in other kingdoms, could be accurately so described. In Abuja, as has been seen above, an important royal slave served as the regular ambassador to Bornu. Other diplomatists might be of humble though free birth, having achieved distinction by their talents, as in the case of Agyei who after beginning life as a salt-carrier on the Volta rose to be second linguist of the Asantehene in the early nineteenth century; being 'always employed in difficult foreign palavers', he was described by Bowdich as the 'Foreign Minister' of Ashanti. Such leading men, writes Kyerematen, were generally given the rule of a village or two but were not recognized as chiefs.[42] An earlier example of this type of official was Matteo Lopez, who led an embassy from Allada to Louis XIV of France. According to the Sieur d'Elbée, his position was equivalent to that of Secretary of State, and he had served several times as ambassador to Benin and Oyo. 'The office of interpreter here', d'Elbée added, 'is very considerable, but the least mistake is as much as their lives are worth.'[43] Similarly, the linguists at Kumasi were required to 'take fetish to be true to each other, and to report faithfully'. Presumably 'new men' such as Lopez acquired wealth from their activities, though presumably too the expenses of an embassy often exceeded the sum granted by governments, despite the hospitality traditionally required from the recipients of an embassy.[44]

Except where the external influences of Islam or Western Europe were strong, there is no trace of even the most vestigial foreign office to serve as a centre for the information and execution of foreign policy. Yet examples of recognized diplomatic staffs can be found in pre-colonial West Africa, and the careers of men like Lopez and Agyei illustrate the professionalism which existed in diplomacy at any rate in the two centuries before the pre-colonial period. Possibly the most highly organized staff was that of the Alafin of Oyo. This was the group of household slaves known as the *Ilari*, also called 'half-heads' from the custom of shaving one side of their heads into which a magical substance was inserted.[45] Numbering some hundreds of both sexes, each male pairing with a female, the Ilari performed multifarious duties, but whereas the junior had administrative and menial tasks within the palace, the senior males acted as a bodyguard to the Alafin and also as his messengers to the outside world. All bore titles, some of which had a significant relation to their calling, as for example: *Oba l'olu* ('the king is supreme'), *Oba ko she tan* ('the king is not ready'), *Kafilegboin* ('Stand fast'), and *Madarikan* ('do not oppose him'),[46] and as credentials they carried fans embroidered in red and green. Another Yoruba court employing 'half-heads' was Ife, where the Oni's messengers were known as the *Emese*, the antiquity of their order being attested by the terra cotta representations found in the town.[47] A somewhat similar body of royal messengers existed in Dahomey, the *Wensangun*. They too were 'half-heads', and it has been supposed that their organization was based on that of Oyo. Travelling in pairs so as to ensure the accuracy and security of the messages which they memorized, these athletic diplomatists were required to bring their oral despatches all the way at the run, and stations were provided on the road from Whydah to Abomey where they could be relieved by relays of others.[48]

The diplomatists of West Africa generally carried credentials or badges of office in such forms as a cane, a baton, a whistle, a fan or a sword. The best known of these are the staffs of Ashanti and Dahomean ambassadors. They were often covered in gold or silver leaf and decorated with symbolic emblems. A favourite Ashanti device for a staff was that of a hand holding an egg, to convey the warning that neither the king nor his representative should press a matter too hard nor treat it too lightly. The staff of the chief linguist of the Asantehene was called *Asempatia*, 'a true account is always brief',[49] Such objects, by extending the power of the ruler beyond his normal reach, were intended to ensure the safe passage of his envoys through alien territory. Similarly, special clothing, a diplomatic uniform in fact, was often worn, like the black caps which, according to Bosman, ensured 'an effectual free pass everywhere' for the Tie-Ties of the Fante. The ambassadors of king Tegbessou of Dahomey to Bahia in 1750 were offered Portuguese clothes by the authorities there, but preferred to appear at their audience with the Viceroy in their own magnificent garb. Ashanti ambassadors were provided by the king with splendid clothes which on their return from a mission they surrendered to 'a sort of public wardrobe'. The Poro officials who enforced their arbitration among the Mende chieftancies went masked as 'devils', impersonating the guardian spirit of their society, while among the Dan of the Ivory Coast and modern Liberia special peace-making masks were worn for palavers – 'large masks with animal-like features and a moveable jaw', a device which modern diplomatists might envy, although not so reassuring as the 'viso sciolto coi pensieri stretti' of the Renaissance ambassador.[50]

The practice of diplomacy requires a measure of immunity for the person and possessions of the diplomatist, and at the least his protection against arbitrary detention.[51] The principle seems to have been clearly recognized in most parts of West Africa in pre-colonial times, especially when the diplomatists carried credentials of the kind just described. Ajisafe in his account of Yoruba law makes a clear statement of diplomatic immunity: 'Embassy between two hostile tribes, countries, or governments is permissible in native law and the ambassador's safety is assured; but he must not act as a spy or in a hostile way...'.[52] Immunity and safe-conduct seem to be rooted in the respect accorded to strangers by nearly all communities other than the most primitive. African hospitality and the honour done to strangers were remarked upon by many of the early European travellers, who themselves received ceremonious and cordial welcomes from the rulers of the countries they visited; equally often they complained of the delays which preceded their reception and the requirement, as Burton put it, to *fare anti-camera*.[53] The same sources show that important African strangers were received with at least equal *empressement* as the white men. In one account the king of Quoja (at Cape Mount) sent to ask visiting envoys to delay their entry to his capital so that a proper welcome could be prepared; the reception then took place in an open hall lined with soldiers.[54] Bosman says that such visits were in general 'accompanied with several Ceremonies', that soldiers were drawn up in the market place and visitors greeted with 'a dismal Military Noise'.[55] The king of Juala (presumably Joal, some seventy miles south of Cape Verde) was so insistent a host, and so much brandy and palm wine were drunk at his receptions, 'that it is much if the king or the envoy comes off sober'.[56] The Ashanti who, Claridge claims, were never known to fire on a flag of truce or to murder an ambassador,[57] were outstanding in their respect for the usages of diplomacy. Among the remotest peoples the laws of hospitality prevailed, yet in some cases immunity was dangerously circumscribed. For the Kagoro of Central Nigeria, for example, guests were sacred on their precincts but when on the road next day they might be murdered, robbed or enslaved.[58]

Nothing corresponding to the fiction of extra-territoriality (a fairly late development in Western diplomacy)[59] protected the diplomatic missions of West Africa. The European forts on the Gold Coast, however, were doubtless thought of by their occupants as places enjoying immunity from local law and were indeed of a similar character to the extra-territorial concessions granted to foreigners in, for example, China and Turkey. In Yorubaland and elsewhere criminals and others who incurred the wrath of the authorities sought safety in recognized santuaries, which in at least one case included the king's palace.[60]

Associated with the treatment of diplomatists and other visitors was the observance of protocol, a subject which only those unacquainted with the history and requirements of international intercourse would dare to dismiss as trivial. As at European courts, there was great variety in the etiquette of the different kingdoms of West Africa, but here again a pattern can be observed. For example, it was usual for kings to converse only indirectly with their visitors and subjects. In Dahomey, the two highest officials, the *Migan* and the *Meu*, spoke respectively as intermediaries for the king to the people and for the people to the king. The Meu, as *chef de protocole*, also took charge of all strangers and supervised public ceremonies. According to Dalzel:

> The King (of Dahomey), and all his subjects, receive strangers with the most remarkable courtesy. Ambassadors, from whatever state, are not put to the necessity of learning the Dahoman etiquette. Every one salutes the Sovereign, according to the fashion practised in his own country.[61]

At the small court of Quoja, special privileges were allowed to the ambassador of the king of Folgias, to whom Quoja was tributary, and also to the Dutch representative who alone was allowed to eat at the king's table.[62] Nor will anyone familiar with the usages of modern West African society be surprised at this emphasis on precedence, a subject closely touching every diplomatist's interest. A vivid, though late, example occurs in Johnson's *History of the Yorubas* when, during negotiations in 1890 by the colonial government at Lagos to end the Yoruba wars, the Alafin of Oyo objected to his representative being given no higher status than the representatives of Ogbomosho and Iwo, towns formerly within his dominion, and refused to send his Ilari to an important meeting. A spokesman remarked that the Governor of Lagos 'ought at any rate to know what is due to a Sovereign or he would not have been selected to represent one'.[63] A pleasing instance of African ingenuity in solving a European dispute over precedence in the early eighteenth century was the occasion when the senior prince at Allada took the quarrelling French and Dutch representatives each by one hand and conducted them simultaneously to their audience with his father the king – though Labat notes that it was the Frenchman who was on the prince's right.[64]

Despite all this, cases of harsh and cruel treatment of ambassadors occurred. Ibn Fartua, for example, reports two occasions on which the troops of Idris Aloma of Bornu killed envoys from the pagan So during wars in the late sixteenth century.[65] The imprisonment of ambassadors by a sixteenth century Oba of Benin has been mentioned above. Bosman refers to the murder in the seventeenth century at 'Great Ardra' – Allada – of an ambassador of the Alafin of Oyo, an act which precipitated the invasion of Allada by an Oyo cavalry force. On several occasions in the eighteenth and nineteenth centuries the Ashanti complained of the ill-treatment accorded to their representatives by the Fante. In 1726, for example, an Ashanti ambassador sent to ask the Fante to withdraw their protection from king Ntsiful of Wassa, who had taken refuge with them, was pelted with stones. This was followed the same year by the first Ashanti attack on the Fante.[66] In 1878 an envoy sent by the ruler of Imesi-Igbodo to assure Ibadan that his town would not join the Ekitiparapo was murdered by his own townspeople on his return home.[67] But these cases seem all to be recognized as exceptional. As elsewhere the career had its hazards, but on the whole ambassadors and foreign visitors enjoyed honour and protection in the kingdoms of West Africa.

The problem of communication with strangers led to the employment in negotiations of those who were skilled in foreign tongues. Ambassadors, like rulers, often had to rely on or chose to speak through interpreters and, as probably in diplomacy elsewhere, there was a tendency for the interpreters to become the negotiators. The spread of Islam made Arabic available as a medium of communication in parts of the West Sudan, and as early as the eleventh century at the court of Ghana the interpreters were Muslims.[68] In Mali the fourteenth-century traveller Ibn Battuta took offence at this indirect method of communication; he also commented on the importance at the court of Dugha, the royal interpreter.[69] The Islamic revolutions of the nineteenth century, and especially the establishment of the Sokoto caliphate, gave a renewed impetus to the speaking and writing of Arabic. Between Europeans and Africans the language used from the sixteenth century onwards was most often Portuguese. Ruy de Pina wrote of the ambassador who in 1486 accompanied d'Aveiro from Ughoton, the port of Benin, to Lisbon, that he 'was a man of good speech' – presumably he learnt some Portuguese on the voyage – 'and natural wisdom'.[70] Roman Catholic missionaries and slave traders spread the use of the language, and this must on occasion have facilitated intercourse between Africans themselves. In the eighteenth century Barbot writes of the interpreter of the king of Sestro speaking 'a little *Lingua Franca* or broken Portuguese'. This language was still used as late as 1854 when the British negotiated the Treaty of Epe with ex-Oba Kosoko of Lagos.[71] When a Bornu prince visited the minor Ottoman court of Tripoli on an embassy in the eighteenth century one of his wives acquired a knowledge of Italian, which was the language used there rather than Arabic.[72] Another reference suggests that in the seventeenth century, at the beginning of Oyo's imperial expansion, the Oyo language ('Yoruba proper') was preferred for some purposes in Allada to the local Aja.[73] Trading contacts spread the use of English and French on the coast in both speech and writing from the eighteenth century onwards and probably earlier,[74] while trade between the forest and the savannah led to the expansion of Hausa.

No trace has yet been found of the development of a specialized vocabulary for the conduct of West African diplomacy, except perhaps in the use of significant titles for such envoys as the Ilari of Oyo and some generic terms for ambassadors.[75] The Poro, and doubtless other societies, did, however, use signs, symbols and allusive language in their internal communications, which served to keep secret their political dealings.[76] On occasion, diplomatic communications took the form of symbolic messages conveyed by objects such as the horse's tail sent to the French emissaries by the Egba in 1884 as a 'sign of alliance'[77] or by cowries and other miscellanea arranged in a significant pattern, as in the congratulatory message sent by the Awujale of Ijebu Ode to Oba Akitoye of Lagos after the latter's restoration in 1851.[78]

Like military operations – of which it is the peaceful counterpart – diplomacy requires the support of intelligence. By its nature, this is dark and hidden, and it is difficult to adduce evidence. Something, however, is known of the Dahomean agents, the *Agbadjigbeto*, instituted by the warrior-king Agaja. Snelgrave writes that Agaja, 'this politick Prince', was able through his spies to discover how much the great men and people of Whydah 'were divided, and that the King was only a Cypher in the Government', information which decided his invasion of that country in 1727.[79] In addition to gathering intelligence, the Agbadjigbeto, who were sent abroad usually in the guise of merchants, were also required to create the impression that Dahomean intentions were peaceful and then, on their return home, to manufacture suitable pretexts for aggression.[80] Another view, not an altogether convincing one, is that the main activities of the Agbadjigbeto amounted not to spying or propaganda in the ordinary senses but to an attempt to undermine a possible enemy by magical means such as burying 'medicine' in his towns.[81] In any event, intelligence was realized to be an important asset in international dealings, and the counterpart of the warmth with

which foreigners were received in West Africa was the suspicion of any seemingly unusual actions or questions by
them.[82] Rather oddly, this suspicion does not usually seem
to have extended to the Muslims who were travelling about
the region in the late eighteenth and nineteenth centuries.
Finally, in a manner unpleasantly reminiscent of the practices in some modern states, the Ashanti found it necessary
to keep a watch on their own servants abroad; if Bowdich
is to be believed, an embassy was usually accompanied by
'A shrewd but mean boy ... in the commonest capacity and
meanest attire' who sent home to Kumasi accounts of the
embassy's dealings.[83]

The conduct of international relations in Africa, like that
of local palavers, was deliberate, patient and tortuous. This
is evident from the accounts of European participants and
observers. 'It is the custom of the Court of Ardrah (Allada)
to make Strangers wait a long Time for an answer', complained the Sieur d'Elbée in 1670.[84] Some of the delays were
due to religious inhibitions; according to Dupuis, there were
only some 150 or 160 days in the Ashanti year which were
considered propitious for diplomatic business. Dupuis concluded that: 'In short, the Ashantees are slow, and I believe,
cautious in cabinet; they are slower, however, in warlike
movements'.[85] Bowdich relates how a discredited Ashanti
ambassador attempted to regain the royal favour by successfully settling a dispute with Wassa, and thereupon fell
further into disgrace, the king explaining that 'no man must
dare to do good out of his own hand, or perhaps he would
find he did bad, as Tando had done, in spoiling a palaver
which he and his great men meant to sleep a long time'[86] –
an African version of Tallyrand's advice to the young men
in his office, 'surtout pas trop de zêle'.

Flattery – for example, the attribution of 'strong names'
– was a familiar diplomatic technique, and the giving of presents was obligatory, both to ease negotiations and as a

token of friendly relations. Dahomey attempted on more than one occasion in the eighteenth century to avert the hostility of Oyo by sending 'great presents', and Allada, threatened by Dahomey, retained the support of Oyo by directing a stream of presents to the Alafin.[87] Gifts were also exchanged traditionally between the Mossi states, between Kumasi and Wagadugu,[88] and between Oyo and other independent states such as Bussa and Nupe. An unusual feature of royal audiences at Benin, reported by Barbot, was that the presents brought by Europeans were not shown to the Oba until after the visitors had withdrawn.[89] Presents took many forms, cloth and coral being favourite items; among the more unusual were the giraffe sent from Bornu to Tunis in 1257, the 'beautiful pair of young leopards in a bamboo cage' provided for the Ashanti embassy of 1821 to take to George IV of England, and the great horn made from an elephant's tusk which the Asantehene sent to the Ya Na of Dagomba about 1830.[90]

The aim of diplomacy being to carry out the policy of a government by means of negotiation, its achievements are usually expressed in either informal understandings or specific treaties, though in pre-colonial West African history it is often impossible to distinguish between these. Such arrangements may be either secret or open. Examples which illustrate the technique of secret, informal understandings may be found in the history of Dahomey, whose conduct of negotiations in the eighteenth century, backed as it was by a Prussian authoritarianism and power, reached a high level of efficiency. According to Dalzel, king Kpengla's envoys were able to bring about a war in 1786 between his enemies at 'New Ardra' (Ajase) and their former allies the Weme.[91] A few months later Kpengla succeeded in similarly separating the Ajase from their protectors at Oyo.[92] The majority of the treaties about which there is evidence were designed to conclude hostilities between states, a topic which will recur in the following chapter, while others (especially those with European visitors) concerned trade. An early treaty of the first kind was that concluded between the Hausa states of Kano and Katsina, *c.* 1650, to end a long series of wars; another was the boundary agreement in the late sixteenth century intended to end Idris Aloma's Kanem wars and which, although apparently ineffective, was 'perhaps the first written border agreement in the history of the Central Sudan'.[93] The peace treaties of 1730 and 1748 between Oyo and Dahomey (the second concluded through the good offices of the director of the Portuguese trading fort at Whydah) are well-known; they aimed at a comprehensive settlement and laid down the annual tribute to be paid to Oyo by Dahomey.[94] Other examples are the alliances concluded by the Fante against the Ashanti in the eighteenth century, the treaty of Jarapanga in *c.* 1830 between Ashanti and the defeated Dagomba,[95] and the anti-Ibadan alliance of the second part of the nineteenth century. The Sokoto caliphate provides an example in the treaty which it made at Togu in 1866/7 with the independent Hausa state of Kebbi, bringing a peace of eight years between these old enemies.[96] In a different category were the small-scale treaties (*ikul*) between sections of the Tiv. A typical ikul imposed meticulous prohibitions on bloodshed and on sexual relations between members of the contracting groups and rules requiring 'almost exaggerated hospitality' when members of one group visited another.[97] A curiosity, and an important case if it can be accepted, was the treaty of neutrality which according to Labat was entered into by the king of Whydah in 1714 with representatives of the French, Portuguese, English, and Dutch, the king refusing to be a party to the hostility between the French and other foreign traders visiting his domains.[98]

Treaties in West Africa were entered into solemnly and were usually regarded as binding and sacred. 'African customary law', writes Elias, 'shares with customary international law acceptance of the principle *pacta servanda sunt* as basis for assurance of a valid world order.'[99] The binding nature of treaties was enforced by the swearing of oaths, which were often formidable undertakings; among the Tiv, for example, they entailed the killing of an elephant and a slave, followed by the preparation of sacred emblems and potions and the mingling and consumption of the blood of the parties.[100] A curious, and now apparently forgotten, ceremony 'of contracting Friendship and Trade' once practised on the Ivory and Gold Coasts was the dropping of saltwater into the eye or rinsing it in the mouth by the parties.[101] Treaties were also confirmed by the giving of hostages, these sometimes being members of royal families, and by dynastic marriages, as in the case above of the exchange of daughters between the Alafin of Oyo and the king of Dahomey in 1730.[102] In appropriate circumstances the parties might resort to devices such as the planting of trees to demarcate a frontier, which concluded a war between Oyo and Benin, probably in the sixteenth century.[103] Yet treaties could be broken in Africa as elsewhere, as was the case with that of 1730 between Oyo and Dahomey. Doubtless Africans were as ready as others to fall back on the specious doctrine that their treaties held an implicit reservation that they remained in force only so long as conditions were unchanged – the *clausula rebus sic stantibus* of European lawyers; according to Bosman, there was a regular and solemn procedure in Guinea by which absolution from an oath could be obtained.[104]

It remains to consider the influence on the system of international relations in pre-colonial West Africa of two alien cultures, those of Islam and of Western Europe. This should provide evidence as to the level of attainment of diplomacy in West Africa by the end of the pre-colonial period and the degree to which it had become fitted to contend with the growing political problems of the time, problems exacerbated by improved communications and more efficient weapons of war.

The introduction of Islamic doctrine into West Africa from the eleventh century onwards was accompanied by that of Islamic law, the two aspects being indeed hardly distinguished. In much of Africa, especially the forest states, this law seems to have been carefully accommodated to the fundamental precepts of the customary law,[105] and with customary law it served to foster unity among different peoples. Rules of warfare received special attention in Islam, and a degree of uniformity was introduced in the Islamized states of the Western Sudan in such matters as the disposal of booty and the treatment of captives – themes to be developed in the following chapter. As regards international relations, Islamic theory defined these in terms of the relations between believers, the members of the community of faith, all equal before God, and non-believers, who were either *dhimmi*, 'people of the Book', such as Jews and Christians, or 'pagans'.[106] 'The concept of the state as an independent political institution is just as alien to Islamic as to African thought', writes Trimingham.[107] In practice, however, Islamic governments were confronted with political questions which divided even members of the faith, and often along national or tribal lines, as, for example, when the Bulala emir rejected the proposal by Idris Aloma, the great sixteenth century Mai of Bornu, to submit the question of three disputed border towns to arbitration by the Shari'a – the Bulala being 'unwilling to follow the Sunna'[108] – or in the contest between Sokoto and Bornu in the first half of the nineteenth century. Yet the Quranic precept to 'be faithful in the keeping of your contracts, for God will require an account of such at your hands', was influential among Moslem states in somewhat the same way as the private law of contract was in the observation of treaties by the nation states of Europe.

In the Muslim world non-resident accredited agents were widely used for the conduct of diplomatic negotiations both within and without the Islamic community. Rules for the behaviour of non-Muslim emissaries in the *dar al-Islam* and of Muslim emissaries in the *dar al-harb* were laid down in detail by the jurists.[109] So far as the international relations of Muslims in pre-colonial West Africa are concerned, however, the only clear examples of embassies seem to be those sent from one Muslim state to another. The earliest known example is probably the embassy which Ibn Khaldun describes as being sent from the ruler of Bornu and Kanem to Tunis in 1257, which brought with it the giraffe.[110] In the fourteenth century Mansa Musa of Mali and his successor Sulayman exchanged embassies with the Marinid sultans of Morocco; so close and friendly was the relationship between the two courts that in 1352 a memorial service was held in Mali for the deceased king of Morocco, Abu'l Hasan.[111] The Moroccan invasion in 1591 of the Songhay empire, which had succeeded to much of the power of Mali, was preceded by a series of diplomatic missions from 1584 onwards, accompanied by intelligence activities, conducted in a thoroughly Bismarckian manner.[112] It is the diplomacy of Bornu, however, for which the best evidence survives. According to a French surgeon writing about 1680, an alliance was established between Mai Dunama and the Ottoman pasha of Tripoli in 1555, followed about a decade later by an exchange of embassies during the reign of Mai Abdullah. Under Mai Idris Aloma, who did much to replace customary law in Bornu by an Islamic juridical system, these contacts with the Ottoman empire expanded, and in 1577 an embassy went to seek a military alliance at Istanbul itself. This seems to have failed, and in 1582 Idris sent an ambassador to al-Mansur, the Sa'dian Sultan of Morocco, with a similar proposal and offering the *bai'a* or oath of allegiance to a Caliph. An embassy to Tripoli in 1578 was reciprocated by the pasha there, who sent gifts of horses and firearms. During the seventeenth century embassies from Bornu continued to visit Tripoli, and there is evidence of occasional diplomatic contact between Bornu and both Tripoli and Istanbul in the nineteenth century.[113]

The most important development in the handling of international relations in pre-colonial West Africa which is associated with Islam is the introduction of literacy, which enabled political records to be kept. Al-Bakri says of Ghana in 1067–8 that the king's interpreters, his treasurer and most of his ministers were Muslims; all of these were presumably literate in Arabic, and the first evidence of writing in Arabic script in the area comes from soon after this time. There is then a gap between the inscriptions at Gao, which indicate communication between Andalusia and the Niger region about 1100, and the writings of Ahmad Baba which reflect traditions going back to the fourteenth century but themselves date from some two centuries later.[114] Arabic literature seems then to have become abundant in the Western Sudan, and includes a number of local chronicles. Leo Africanus noted that in Timbuktu books brought from Barbary were 'sold for more money than any other merchandise'.[115] Typical of the period before the Fulani jihad were the works known as 'Mirrors for Princes', of which the advice of al-Maghili to the king of Kano on *The Obligations of Princes* is an example. Unfortunately little which would throw light on the day-to-day conduct of diplomacy seems to have survived from before the nineteenth century. Considerable information then becomes available about the inner working of the Sokoto Caliphate, especially the role of the Vizier and his chancery.[116]

Among the forest states, it is clear that by the time of the visits to Ashanti of Bowdich and Dupuis, in 1817 and

1821 respectively, a number of literate 'Moors' were resident in the capital. They were apparently mainly engaged in writing magical formulas for amulets, but according to Bowdich a chancery had also been set up by them for recording 'the greater political events'.[117] These Moors – by whom are meant Muslims mainly from the Islamized countries of Gonja, Dagomba and Mamprussi to the immediate north – had settled in Kumasi during the reign of the Asantehene then on the throne, Osei Bonsu, having been previously 'only visitors'.[118] The 'Moorish secretary' referred to by Bowdich had also 'resided some time at Hio', that is, at Oyo.[119] This is insufficient to suggest that records were kept by the Alafin's government, but there is other evidence of some degree of literacy in Oyo in a report by a French trader that in or about 1787 he met ambassadors from Oyo on the Benin river who knew how to write and calculate in Arabic. These ambassadors seem likely to have been northerners from the Sudan, perhaps Hausa, in the Alafin's employ rather than natives of Oyo.[120] Similarly, the king of the coastal state of Porto Novo sent a group of four Muslims, who again were probably traders (or possibly wandering Mallams) from the Sudan, as his ambassadors to Dahomey in 1864.[121]

Europeans arrived on the Guinea coast some three hundred years after the Muslims had made their first converts in the savannah, and from the end of the fifteenth century onwards that part of West Africa which was within reach of the coast was in almost continuous contact with Western Europe. A part of this contact consisted in the small but steady flow of African visitors both to the European settlements in Africa and to Europe itself, most of whom were charged with official business and can be termed, as they usually were by contemporaries, 'ambassadors'. The first visitor of this kind – as opposed to the few unfortunates who were taken to Europe against their will as curiosities – seems to have been the chief of Ughoton, who accepted d'Aveiro's invitation to accompany him on his return to Lisbon after his exploratory visit to Benin in 1486. In 1514 the Oba of Benin despatched a second embassy to Lisbon, this time on his own initiative. The objects of the mission were several: to complain about the conduct of some of the Europeans trading on the Benin river, to answer European complaints about the trading practices of Benin, to discuss the provision of Christian missionaries, and to ask for firearms. The two ambassadors, called 'Dom Jorge' and 'Dom Antonio' by the Portuguese, were probably themselves already Christians. Their embassy, which had little success, was followed by another in 1516 and yet another in or about 1540.[122] Meanwhile, in 1513 an embassy from the king of Congo had reached Lisbon, whence it proceeded in company with three Portuguese envoys – all men of rank who were fully accredited – to Rome in order to make an act of obedience to the new Pope.[123]

In the seventeenth century the small kingdom of Allada provides two interesting examples of embassies to Europe. The first was sent by king Toxonu to Philip IV of Spain in 1658 to protest against the establishment of a French mission post at Whydah, which (in Akinjogbin's interpretation) ran counter to Alladan ambitions that European activities should be concentrated in their kingdom. The second embassy, of which an account was compiled by the Sieur d'Elbée who accompanied it, was sent by Toxonu's successor, king Tezifon, to France in 1670 to discuss trading difficulties at Allada. The ambassador, Matteo Lopez, had the honour of being received at the Tuileries by both Louis XIV and his queen, the latter conversing in Spanish while Lopez replied in Portuguese.[124] On the Gold Coast, envoys from the states there paid frequent visits to the headquarters of the European traders and occasionally to Europe. In 1612 two representatives of the king of Asebu carried out a successful mission to Holland where they won Dutch support against the Portuguese, and in the 1660s an ambassador was sent to France by the Komenda to ask Louis XIV to direct his traders towards their state.[125] Other examples of African–European embassies may be found in the history of Dahomey, which despatched four missions to the Portuguese at Bahia and Lisbon between 1750 and 1811.[126]

European contact with West Africa was maintained in pre-colonial times by missionaries and traders rather than by officials, and only a few instances can be adduced of the sending of diplomatic representatives from Europe to Africa. The earliest seem to have been the two ambassadors sent from Portugal to the king of Congo on separate occasions at the end of the fifteenth century, Rui de Sousa and Simão da Silva.[127] The Portuguese also sent an embassy to the Fula king in Upper Guinea at the end of the fifteenth century, and two embassies to Mali, one about 1490 and the other in 1534.[128] Two prominent examples of much later date are the British missions to Ashanti led by Edward Bowdich and Joseph Dupuis in 1817 and 1821 respectively.[129] The latter, a member of the consular service who had filled posts in North Africa, was scornful of the blunders of his less professional predecessor who, he alleged, was 'unaccountably deficient in a knowledge of the usages of courts'.[130] More frequent were the missions from Europeans settled on the coast to neighbouring African states. The Dutch and English, for example, sent either representatives from their own nations or local African employees to Ashanti. These began with David van Nyendaal's visit to Kumasi in 1701, intended to take advantage for the Dutch of the trading opportunity created by the Ashanti's overthrow of Denkyira at the battle of Feyiase and the opening of an Ashanti route to the guns, ammunition and other trade goods available on the coast. The government of the British colony of Sierra Leone in the mid-nineteenth century kept in touch with the hinterland by employing local officials as 'trek diplomats', of whom the best known was Thomas Lawson.[131]

The object of these missions was usually to obtain trading concessions, though another common motive was to ask for military aid. Numerous agreements were made between the Dutch and the states of the Gold Coast in the seventeenth century. An informal trading arrangement with Asebu as early as 1598 was replaced by a definite agreement in 1656. The Dutch also made a military agreement with Fante in 1624, trading agreements with Axim, Accra, Jabi and Komenda in or about 1642, and a reciprocal military agreement with Fetu in 1688.[132] Another example of a formal commercial agreement was that signed between Benin and the Dutch West India Company in 1715, 'after endless delays and no less expense and labour', as the Dutch negotiator wrote.[133] In 1730 Edward Deane, director of the Royal African Company's fort at Whydah, made a far-reaching agreement with Agaja of Dahomey in which the king apparently agreed to accept responsibility for the safety of all Europeans in his kingdom and to co-operate fully with the European slave traders, while Deane on behalf of his company recognized Agaja's title as ruler of the recently-conquered Whydah and agreed to concentrate trading activities there.[134] Another interesting treaty was that signed by the Fante states in 1753 with the English at Cape Coast Castle, agreeing to prohibit any French settlement in their territories. Unlike this, the treaties signed by Bowdich and Dupuis with Ashanti proved short-lived. The Ashanti were alert to any threat to their independence, as were other African states, such as Bornu and Sokoto in their treaties with Barth, the British emissary, in 1852 and 1853, and Sokoto and Gwandu in their treaties with the National African Company in 1885.[135] A diplomatic curiosity is the possibly forged treaty obtained by a German expedition with Gwandu in 1895.[136]

The arrival of Europeans on the West Coast coincided with, and indeed partly resulted from, the emergence in Western Europe of the modern territorial state, and this in turn, as has been noted above, was the progenitor of permanent diplomacy in the fifteenth century. The new system spread throughout Europe during the next three hundred years, until it was adopted by the Ottoman empire at the end of the eighteenth century.[137] No effort was made during this time to introduce the system into West Africa, though the forts of the European trading companies had something of the character of standing embassies and their directors were sometimes called upon to perform semi-diplomatic functions. Then the abolition of the slave trade in the early nineteenth century caused the governments concerned to look for means of encouraging legitimate trade and repressing attempts to revive slaving. One of these means proved to be permanent representation. Perhaps the only positive result of Bowdich's treaty in 1817 with the Asantehene was the establishment of a British resident at Kumasi. In the event he was soon recalled, but years later, in 1853 Asantehene Kwaku Dua I asked for a resident to be stationed again in Kumasi and the British assigned an educated African, George Musgrove, to the post. An important step was taken by the British government in founding the consulate of the Bights of Benin and Biafra in 1849, a post whose holders, far from confining themselves to consular duties, behaved as ambassadors and even as embryonic governors.

One general result of these diplomatic relations between West Africans and both the Muslims and the Europeans was the enhanced influence of the literate élite. In some cases these men belonged to royal or noble families. At least equally often, it seems, they were men of humble birth, 'new men', like the diplomatists of Wolsey or Francis I or Charles

V. An example of these latter was Agyei, the Ashanti linguist referred to above. Another was Pierre Tamata, the powerful French-educated Hausa who was secretary to the king of Porto Novo in the late eighteenth century, described by a French sea captain as having the malice of a monkey, greater cunning than the fox, and the greed of an eagle – he was an illustration, Labarthe thought, of the danger of sending negros to college in France.[138]

III. West African wars: their causes and character

'War', a word used perhaps as often as any other by historians, is an interpretative term whose meaning, rarely examined, requires brief discussion. Most simply it is the opposite of peace, but then 'peace' too – probably even more than war – is an imprecise condition. In this book 'war' is taken to imply 'a state of open hostility between nations', which is one of the alternative definitions given by the *Oxford English Dictionary*. Skirmishes, battles, and even campaigns, are not wars but incidents comprised within a war. A war may, though rarely, be decided by a single battle, but that battle is something less than the total state of hostility within which it takes place. Again, there is a formality about war which, varying according to the sophistication of the protagonists, confers a degree of legality; it is, for example, a condition recognized by international law. Apart from the case of the complete collapse of one of the contestants, a war, even when 'undeclared', must usually be ended by agreement, more or less formal, between the belligerents to make peace or at least to observe an armistice; exceptionally, it may die away through their exhaustion or inertia.

The application of this word 'war' to West African hostilities involves no difficulty, though some of the languages of the region are even less precise than English in their usage of the term, or what must do duty for the term; in some languages a single word can imply a war, a battle, or just a skirmish. European observers have often used such limited terms as 'raid', 'expedition', or 'campaign' to describe the wars of West Africa. Certainly, up to the mid-nineteenth century, military activity there seems almost always to have been confined to the dry season, lasting from about October to May in the savannah and from about November to March on the coast and in the forest. The coming of the rains might halt a successful campaign in mid-career, as in the case of that conducted in 1765 by Ashanti and Fante in temporary coalition against Akyem.[1] Such cessations of hostilities were not due only, and perhaps not primarily, to the difficult conditions created by the heavy downpours of the tropical climate, but were occasioned also by the need for the soldiers to return to work on their farms. Then, as so often in Europe, the completion of the harvest was a signal for the renewal of hostilities.[2] Other limitations on warfare were religious, certain days or seasons being deemed unpropitious. For Moslems, fighting was discouraged, if not actually prohibited, during the three (or four, according to some authorities) sacred months, but this had little practical effect; Usman dan Fodio, overriding the precept, told his followers to 'Slay the idolaters whenever you find them.'[3] In such respects, however, Islam was more flexible than many other religions, as for example that of Ashanti whose imposition of very numerous holy days was noted in the last chapter.

Thus it was climatic conditions which most often regulated hostilities in West Africa, and campaigns were in general regular dry season events, like Idris Aloma of Bornu's wars in the sixteenth century against the Bulala in Kanem or the annual marches of Oyo against Dahomey in the eighteenth century, when the Dahomeans habitually retired to 'their fastnesses and woods'.[4] In such cases a single campaign might be described in political terms as a war. Yet a state of general hostility between adversaries might indeed last, though with only intermittent action, for several years, especially when, as Bosman puts it, the combatants were 'two Despotical Kings, who have their Subjects entirely at their Command'. In some such cases the armies would 'lye a whole Year incampt against each other' without any action taking place other than a few skirmishes, and breaking off entirely during the rains.[5] Other wars were of a nature which might almost be characterized as ceremonial, like the campaigns which a new ruler of the Hausa state of Maradi was required to mount within two weeks of his accession – though the objective of the Maradi was real enough, being the recapture of their old capital, Katsina.[6] This pattern of mainly short wars, taking place within one campaign, varied by some longer but intermittent struggles between the richer and more highly organized states, seems to have prevailed throughout West Africa in the seventeenth and eighteenth centuries, and was probably much the same in earlier times. Towards the end of the period, however, when professional armies largely displaced militia and were dependent on expensive imported weapons and ammunition, wars tended to be longer in duration, as in the cases of the Ijaye War of 1860–5 and the Sixteen Years War of 1877–93, both in Yorubaland. Such wars took less account of the seasons; the Egba, for example, marched to the aid of the Ijaye in June 1860 at the wettest time of the year. The troops remained in the field, though accommodated in camps which increasingly came to resemble permanent villages, and continued to harass their opponents throughout the year.

The predominance of seasonal campaigning may suggest that West African wars can be explained largely and simply as a reflection of African aggressiveness. The negro, some writers have held, is the most warlike of the races of men,[7] and for him war, the satisfaction of his psychic needs, was a way of life. This may have been true to some extent of those societies which had not developed central institutions, yet even here the affrays and long-lasting feuds of such

peoples as the Kagoro and the Ibo had their origins in
specific and ascertainable grievances and rivalries over such
matters as disputed boundaries, ill-treated travellers, stolen
women, or hunting accidents.[8] Generalizations about
'primitive' war and 'primitive' peoples have only a restricted
value for the interpretation of the history of stateless societies. They have none at all where states of the kind which
covered much of pre-colonial West Africa are concerned.
The causes of war there were as multifarious as elsewhere,
rooted in the interests of individuals and of groups and in
converging sequences of events.[9]

A fundamental cause of most West African wars – and,
indeed, the most prevalent cause of wars in any part of the
world – was the desire of the more vigorous societies for
territorial expansion and to exercise a measure of physical
control over their neighbours. This, according to Hinsley,[10]
was the dominating characteristic of European political
history down to the eighteenth century and is far from
having disappeared as a motive. It accounts at least in part
for most of the wars which will be noticed in this book:
for the strife which led to the displacement in turn of each
of the great empires of the medieval West Sudan from
Ghana onwards, for example, for the Kanem wars of Idris
Aloma, designed to regain possession of the original home-
land of the Bornu nation, and for the frequent Oyo expeditions against Dahomey in the eighteenth century. But in
West Africa territorial aggrandizement was by no means
the necessary or normal sequel to victory in such wars; perhaps more often a loose tributary relationship, such as Oyo
imposed on Dahomey and others of her neighbours, was
preferred to annexation and the extension of state boundaries.

The acquisition of wealth was an ambition of equal, or
nearly equal, importance to that of power over other
peoples. This might entail the occupation of farmland, and

thus was bound up with territorial expansion. Another common aim was the exaction from the conquered of tribute, which in suitable circumstances was preferable though akin to the taking of booty. Access to trade and the control of trade routes were similarly strong motives for war. One object of Idris Aloma's expedition to the Kauwar oasis in the late sixteenth century seems likely to have been to assure the supply to Bornu from that region of natron and to keep open trade with Fezzan, Egypt and the Mediterranean coast. Other interior peoples, nearer to the Atlantic, were anxious to enter into direct relations with the European and other merchants on the coast from whom arms and ammunition were obtained. Trade rivalries were a principal source of hostility between the Fante states themselves and between the Fante and the Ashanti, and expansion to the coast was an important objective for the Dahomeans in their wars of the late eighteenth and the nineteenth centuries and the Ashanti in the eighteenth and nineteenth centuries.[11] Other, and cruder, motives for war were the plunder of moveable property and, of far greater importance, the taking of captives.

The importance to be attached to the capture of slave material as a cause of West African wars is a topic of considerable interest for historians. The abolitionists of the eighteenth century all placed great stress on the connexion between the slave trade and the incidence of war, and Turney-High, a modern sociologist, has written that in Africa 'serious farming, a fortunate geography, and a teeming population produced slavery, and slavery meant war'.[12] There is support for these views in the writings of contemporary observers of pre-colonial West Africa. Leo Africanus noted the dependence of the supply of the larger horses used by the cavalry of Songhay and Bornu in the early sixteenth century upon the slaves of the Barbary merchants.[13] Under Islamic rules of war the enslavement of captives was permitted,[14] and the capture of 'pagans' as slave material became a prime object of Muslim wars. The raids carried out by Islamic states such as Bornu and Sokoto against their less sophisticated neighbours reaped a rich harvest of slaves, and the conclusion cannot be resisted that in such cases religion furnished only a pretext for war. Denham writes that though the expedition against the Munga which he witnessed in 1823 was ostensibly mounted because this semi-pagan people were 'kaffering, and not saying their prayers – the dogs!', the real motive was the capture of slaves.[15] The Atlantic trade was even more demanding. Dapper, for example, noted that large numbers of war captives and 'criminals' sent down by the Oyo were bought by the Dutch and Portuguese at 'Little Ardra' (Porto Novo), while Barbot claims that enslavement was the fate of all those who were taken prisoner of war.[16] In return for slaves, the Europeans supplied the coveted fire-arms, ammunition and other weapons.

After the 1820s the Yoruba, who until then had been bene-ficiaries rather than victims of the slave trade, provided much of the raw material for the Atlantic exports. This was largely a consequence of the numerous wars and the political instability which afflicted their country for the most of the century. The degree to which this instability was brought about by the demands of the slave trade – the identification of cause and effect – is controversial. According to Ajayi, 'the role of trade generally and the slave trade in particular' in the Yoruba wars 'has been much exaggerated' and he has drawn attention to the important political causes – especially the collapse of the Oyo empire and its system of government – which lay behind them.[17] Hopkins, on the other hand, has tried to show that 'In their final stages' – an important limitation to his argument – 'the wars were the climax of a prolonged effort to carry into the new age the political supremacy which was so firmly based on the old economy of the Atlantic slave trade',[18] a view which produced a vigorous rebuttal from Ajayi and a colleague.[19] But such debate about the motives of humans and their institutions can rarely be satisfactorily resolved. Clearly, numerous expeditions and campaigns conducted in nineteenth-century West Africa had as a principal object the taking of captives destined for the slave market – even more so, the kinds of raids in Yorubaland of which the boy later to become Bishop Samuel Ajayi Crowther was a victim.[20] Again, slave catching was a prominent feature of such notable upheavals as the Ijaye War and the sixteen year long war of the Ekitiparapo.[21] Equally clearly, the more important wars had other identifiable causes, political, economic and occasionally ideological, of equal and often greater importance than the provision of raw material for the trade in slaves. This trade exacerbated political relations, prolonged wars, and contributed greatly to the political instability of West Africa, especially the Slave Coast and its hinterland, for many years, but apart from those limited operations correctly designated 'raids' by contemporary observers, it seems not to have been a fundamental cause of war itself.

With the important exception of the Islamic revolutions which took place in West and Central Sudan, ideologies seem to have played little part in bringing about the wars of West Africa. Indigenous religions were apparently not interested in proselytization beyond their ethnic boundaries, and although it might be argued that the absorption of one people by another more competent in warfare was an extention of the latter's political doctrine, this would be a thesis practically impossible to prove. But one important West African case of conflict motivated by indigenous religion is the wars of the Dahomeans in the eighteenth and nineteenth centuries, a principal object of which was to take captives who were subsequently to be sacrificed to the ancestors of the ruling dynasty; this ritual requirement was one on whose satisfaction the well-being and the very maintenance of the kingship were held to depend.[22]

Christianity came to West Africa peaceably, and until the missionary impetus of the nineteenth century got under way its influence was on a very small scale. The spread of Islam in the West and Central Sudan and its tentative penetration of the coastal regions were largely the work of itinerant traders; neither the Almoravid conquest of Ghana in 1076–7 nor the Moroccan invasion of Songhay at the end of the sixteenth century were motivated by religious ideals. Nevertheless, for the Muslim, according to the *hadith*, 'Holy War is the peak of religion', and from the seventeenth century onwards, beginning in Mauretania and Senegal, there occurred a series of jihads.[23] The most prominent was that led by Usman dan Fodio and his sons which between 1804 and about 1831 created a vast empire in Northern Nigeria on the ruins of the Hausa and other states and also challenged the ancient Islamic state of Bornu. In the west there followed the holy war of Seku Ahmadu which founded the short-lived imamate of Macina (*c.* 1818–62). This was overthrown and succeeded by the empire of Al-Hajj Umar, which in its turn was destroyed within thirty years by the French. Numerous later wars – notably those fought by Samori – aspired similarly to the role and title of jihad. That most of these wars were fought in their opening, and most important, stages against states which had for long accepted Islamic beliefs does not invalidate a description of them as being conducted primarily for ideological reasons since, as Al-Maghili had explained to Askia Muhhamad of Songhay in the fifteenth century, the jihad against lukewarm Muslims and backsliders was even more important than the war against unbelievers. Once a properly constituted Islamic state had been established, however, these Muslim conquerors aimed primarily at ensuring the provision of tribute and levies rather than at proselytization or political domination.[24]

Though indigenous religion, like ideology generally, was only exceptionally of importance as a political force in pre-colonial West Africa, appeal to the supernatural was prominent at every stage in warfare. Priestly diviners were almost always consulted before a decision was taken to make war, though, as Barbot remarks, they were unlikely to advise its prosecution unless there was an apparent preponderance of strength over the putative enemy.[25] The next move might be, as among the Dahomeans, to send out spies to bury magical substances in the enemy country.[26] Before the army moved off, sacrifices were made to the war god or war standard, and weapons were smeared with magical potions; the Ibibio priests also rubbed an ointment on their warriors' bodies to render them 'invisible'.[27] Superstitions and tabus abounded; Ibibio warriors were forbidden in war to take 'soft food, and if a group of Ibo raiders met a certain kind of snake on their path, they desisted from their operations.[28] Charms and amulets were in demand by both pagan and Muslim soldiers, usually being stitched onto the garments worn in battle and affixed by cavalrymen to their horses' necks; Muslims were favourite providers of various kinds of these for use by pagan armies such as those of Gonja and Ashanti.[29] Sacred objects were brought into battle, like the ancestral black stools of the Ashanti or the ritual stool used by Balogun Ibikunle, the Ibadan general,[30] while drums were beaten at least as much for their religious or magical powers as for making signals or rallying the troops. Among Muslims there was frequent resort to prayer. In addition to the regular prayer required daily of all the faithful, and which in the field underpinned military discipline, special instrumental prayers were said at appropriate stages in the action.[31] The frequent presence of clerics with the armies engaged in jihad enabled regular instruction to be given to the soldiers on the principles of their faith and on the special rewards awaiting those killed in holy warfare. The Bornu, apparently even after their profession of Islam, depended for victory in war on their mysterious Mune, 'a certain thing wrapped up and hidden away', until Mai Dunama Dibbalemi impiously opened it.[32] Considerable reliance was placed on appeals to the supernatural, as when, for example, during an attack by the Dahomeans on their capital, the Whydah installed their sacred snake on the banks of the river which lay across the enemy's path, and neglected more obvious means of opposing the crossing of this obstacle.[33] Finally, when battle was done and captives and trophies (such as the heads of slain enemies) secured, cleansing ceremonies by the participants might be required. These often included rituals to exorcize the ghosts of the dead, such as the licking by the Ibo warrior of enemy blood from his knife.[34]

Conflicting views have been advanced about the determination and zeal with which West African wars were fought. On the one hand it had been claimed that these wars of the negroes were exceptionally hard and cruel, very costly in lives, and on the other that they were conducted with such mildness and even timidity that casualties were ridiculously low so that in this respect they were akin to the allegedly unreal wars of the Renaissance in Italy (commenced without fear, continued without danger, and concluded without loss', as Machiavelli claims – unreliably). There is evidence to support both views. Idris Aloma's victory over the Bulala, for example, is said to have cost the Bornu only one man dead and four wounded. Bosman, for example, writing in general terms about negro warfare in West Africa, claims that few were ever killed in battle, 'they are so timorous', while Atkins says of the inhabitants of the Gold Coast that 'it is a bloody Battle among them, when half a dozen of a side are knocked down'. Others again single out the poor performance in war of the Whydah and Allada soldiers; though the Whydah could bring (it is said) 200,000 men into the field, they were 'weak and heartless', more inclined to trading and husbandry than war, and only prepared to fight when on the defensive.[35] Engagements might be deferred or avoided altogether through the nomination of champions as duellists.[36] Though Dapper praises the courage of the Benin army, his opinion in this was contradicted by both Bosman and Barbot; the latter explains the extent of the Benin empire by saying that the nations to the north and west were even less soldierly.[37] Bosman makes an exception of the Aquambo on the Gold Coast who were 'Haughty, Arrogant and War-like', a terror to their neighbours.[38] For Muslims, though the promises of their religion sustained the timid and made the brave braver, theologians still found it necessary to lay down clear rules of conduct in battle: for example, an army was forbidden under penalty of grave sin to retreat if the enemy were less than double the number of the faithful, and if the Muslim army was over 12,000 in strength, united and armed, flight was forbidden under all circumstances.[39]

Yet casualties could on occasion be heavy, testifying to the reality of West African wars and the determination with which they were fought. Bosman's statement that 100,000 fell in the two battles in which the Ashanti overcame Denkyira in 1701 cannot be accepted at its face value, but this was doubtless a bloody affair and much was at stake. Again, after the overrunning of Allada in 1724 by the Dahomeans, according to the account of the Englishman, Bullfinch Lambe, who was captured there, 'there was scarce any stirring for Bodies without Heads; and had it rained Blood, it could not have lain thicker on the Ground'. The wars of the Yoruba in the nineteenth century, both among themselves and with their neighbours, also claimed a heavy loss of life, for which there is evidence from missionary and other outside observers. Again, after the battle of Tsuntua, in which the jihadists of Usman dan Fodio were defeated by the Hausa of Gobir, assisted by Tuareg allies, some 3,000 dead were counted.[40]

Thus no generalization seems applicable to West African wars in this respect, so wide is the field in both time and space, and it can only be concluded, platitudinously, that the cost in human lives and suffering differed greatly according to circumstances. Three general impressions, however, do persist, difficult to substantiate but for which the wars of the eighteenth and nineteenth centuries provide some evidence. The first is that non-combatants in the theatres of war often suffered to a greater extent than those who took an active part in the fighting; if they returned safely from hiding places in the bush, in caves, or on bare hills, they were likely to find their farms devastated, and if they were taken by the enemy they were as liable as the soldiers to be slain, ill-treated or enslaved. Secondly, it appears that during the nineteenth century there was a widespread increase in the scale of West African warfare and in the bitterness of the fighting. It may be objected that evidence from previous centuries is insufficient to allow any comparison. Yet, to take the most prominent examples, the holy wars of the Muslims of the Central and West Sudan, the long and bloody aftermath of the collapse of Oyo, the Ashanti break-through to the coast, and the consolidation of Dahomean power, all were political and military events of an unusual magnitude and importance, while the growing supply of firearms, and of firearms which were far more lethal than before, was making war a more drastic, efficient and terrible means of achieving political aims than had ever been known in the region. Finally, it seems clear – though the evidence is only negative – that, violent and disagreeable as West African wars must have been, the area of devastation was narrow compared to that created by modern European war. Doubtless a large proportion of the population, living far from the field of battle and the track of armies, continued peacefully to follow their occupations in their villages and on their farms, undisturbed by and probably ignorant of the great events taking place around them.

As elsewhere, the warfare of West Africa was mitigated by the establishment of conventions. Of these the formal declaration of war was perhaps the most important, giving an enemy time to prepare for an attack and an opportunity for parleying and for sending women, children and the elderly to safety. Among the centralized states, the decision to make war was a deliberate one, taken usually not by the ruler alone but with the advice of his council, and solemnly promulgated. Before hostilities opened, the Asantehene, as has been seen above (p. 18) placed himself on the *hwedom* ('facing the enemy') in order to pronounce a declaration of war.[41] Such deliberation amounted to a conscious sacrifice of the advantage of surprise in the attack. The Fante, according to Labat, sent a herald to the enemy to declare war and to make proposals for the time and place of battle, while in a war between the Dahomeans and the 'old Whydahs' with their Popo allies in 1743 the captains are said to have first 'held a dispassionate convention at the head of their troops' and drunk a toast together before the opening of hostilities.[42] Such chivalry was perhaps not-typical, but its occurrence was not confined to the more highly developed states. The Gannawarri of Northern Nigeria, for example, prescribed a three-day delay between a quarrel with and an attack upon a neighbour – called in Hausa the *wasan wuka* or 'knife sharpening time'. The Ibo, Ijo and other peoples of South-eastern Nigeria also gave notice to their enemies of an intention to go to war with them, laying plantain leaves and piles of powder and shot on the paths and arranging for the times and places of battle – though these battles, usually taking place on the local boundaries, were little more than bush tournaments.[43]

In Western Europe conventions mitigating the cruelties of war developed to the point where it becomes possible to speak of the 'rules of war'. These rules, which formed an integral part of early international law, indicated the conduct expected of victors and the rights which remained to the vanquished. Though only a few traces of such advanced doctrine can be detected in West Africa, conventions of various kinds were accepted there as applying to warfare within certain groups, in addition to the formal declarations of war referred to above. Thus, for example, in small-scale Ibo wars over such matters as forfeited bride-price, the use of firearms was prohibited, though this, predictably, was not always effective.[44] Islamic doctrine about the rights and wrongs of warfare was more highly developed but offered little protection to those who were not of the faith. According to most Muslim commentators, unnecessary shedding of blood, the seizure of an excessive amount of property, and the molestation of non-combatants were all forbidden, while some also deprecated the use of poisoned arrows. These merciful injunctions were not, however, repeated by Sheikh Usman dan Fodio in his handbook for those taking part in the jihad, the *Wujub*, which exhorts the warrior to 'Wage total war on the idolators'.[45]

Arrangements for the treatment of the wounded in war, whether friend or foe, seem to have been accorded little importance, a result perhaps more of the lack of medical knowledge than of indifference. The picture is not wholly dark. After defeating a Songhay army in the mid-sixteenth century, the people of Katsina are said to have tended, and later released, their wounded enemy prisoners;[46] their care for their own wounded was presumably correspondingly greater. Bosman writes that 'The Natives are very much to be pitied, that being shot, cut or otherwise wounded in their Wars, they neither know nor have any other way of care than by green Plants'; these plants were boiled and then applied as fomentations, proving 'effectual in some cases'. The Dagomba had various concoctions, usually said to have magical properties, for the treatment of arrow and gunshot wounds, and elderly men followed on in the rear of their armies to apply such remedies. The Jukun had similar arrangements. More alarmingly, the Ibo treated wounds by means of sympathetic magic, tending a different part of the body from that affected.[47] As the nineteenth-century evidence shows – again, the Yoruba wars provide much material – many of the wounded must have perished, either from their injuries or at the hands of their enemies. Their flesh sometimes provided a meal for the victors, their skulls served as drinking cups or ornamented house walls, while the bodies of those who died on the field remained unburied, a prey for the vultures and other wild beasts.[48]

Captives were always regarded in West Africa as part of the booty of war. On the Gold Coast, as probably in most of the non-Islamic kingdoms, they belonged in theory to the kings, but in practice many were distributed to chiefs as rewards for bravery and service.[49] The teaching of the Islamic jurists about captives varied, but always enjoined good treatment for those who were enslaved; women were not to be separated from their husbands nor children from their parents – though once again practice usually fell far short of precept. Moreover Muslim rulers such as Idris Aloma not only made war on professing Muslims but also took captives from them as slaves.[50] As has been seen above, many – almost certainly, from the seventeenth century onwards, most – captives were retained or sold as slaves, and then either exported or directed into one or other form of domestic slavery. In either case, it must be remembered that enslavement represented a considerable mitigation of the harshness of war, for slavery of any kind was preferable

to violent and often hideous death and also probably preferable to imprisonment – which was an alternative rarely
offered to any but the most important captives. Moreover,
many of the captives, especially those taken in the wars
of Yorubaland in the later nineteenth century, were con-
scripted not for plantation slavery overseas but for the relatively mild form of domestic slavery which prevailed in West
Africa.

For many captives a far worse fate than slavery was
reserved, that of death, sometimes swift, sometimes after
prolonged and deliberately inflicted suffering. Idris Aloma
of Bornu, when making war on his pagan neighbours around
Lake Chad, habitually caused all male prisoners to be executed, while the women and children were taken captive –
the sparing of these latter being an example of one of the
earliest ways in which the rigours of war came to be softened.
Three hundred years later the armies of Bornu were equally
merciless, Barth observing after a skirmish with the Musgu
in 1851 that their Bornu captors caused 170 of between 500
and 1000 prisoners to bleed to death by severing one leg
of each. In Dahomey, as already noted, large numbers of
prisoners – nine-tenths of those captured in war, according
to Dalzel – were ritually slaughtered for religious purposes;
on one occasion some 4,000 Whydah perished in this way.
Prisoners of the Ashanti could also in many cases expect
to be put to death, although exceptions were made for those
who could claim to be Muslims, even though Islam was not
the religion of the state.[51]

Death and enslavement were the usual but not the only
possible fates of captives. Some might find themselves con-
scripted into the armies of their captors, as happened to
a number of Akwamu taken by the Akyen in 1730.[52] A more
fortunate few – almost always, the nobler and richer – were
kept as hostages or ransomed. Idris Aloma, for example,
accepted a ransom for two Bulala nobles, while Prince

Tegbesu of Dahomey spent some years as a hostage in Oyo, and after their victory over Dagomba about 1830 the Ashanti took two chiefs from the Ya Na's bodyguard to Kumasi, where they were accorded the status of ambassadors and later allowed to return home.[53] From the meagreness of the evidence about the payment of ransom before the nineteenth century, it appears that usually the demands were set too high to be met. But in the Yoruba wars of the nineteenth century it was quite usual for ransoms to be paid, and exchanges of prominent prisoners also took place.[54]

West African wars were not intended only to pay their way politically, but also to contribute substantially to the economy of the state. Though the taking and enslavement of captives was from the sixteenth century the chief means by which this was achieved, other forms of booty were also important. Portable plunder was the most convenient and prized, especially in circumstances where the victor's resources did not allow him to undertake the administration of more territory. Captured horses were greatly valued, and the weapons of the vanquished could also be put to use by the victors. Booty of this and other kinds was often sold (like captives) and more arms purchased with the proceeds, especially firearms, thus closing the vicious circle of aggression. The allocation of booty in non-Islamic armies varied, but the Jukun practice, whereby up to half of the spoils of war, theoretically all the property of the king, were distributed to the warriors, may have been typical of centralized states, especially those where the ruler himself did not usually take the field.[55]

Islamic teaching on the disposal of booty was clear and detailed. Spoils of war might be taken forcibly only from non-Muslims. They could belong only to those who took part in the war, and were to be divided only after a battle had been won. One-fifth of these spoils, the *khums*, belonged to the ruler of the faithful, the Imam or Caliph, and were intended to meet expenditure on public utilities. The remaining four-fifths was to be divided in various proportions, some teachers, including Usman dan Fodio, allotting three parts to a cavalryman (one part attributed to himself and two to his horse) against only one part for a foot soldier. Slave troops had no legal entitlement to a share, but the enrichment of slaves by this means was far from unknown. There was some disagreement as to whether the division of spoils could or should take place in enemy territory, the dar al-harb, or should await return to the land of Islam.[56] In practice these rules seem to have been interpreted with elasticity. In the Sokoto empire up to one-half of the captives seem to have been sent to the Caliph. There were rather different arrangements again in Zazzau where the Fulani emir and his general, the Madaki, each took a half share from which they rewarded their officials and troops. Barth's account of the division of spoils by the Bornu after battle speaks on one-third being allotted to the Vizier as commander-in-chief; the division had to be made before the army left hostile territory. Other features of the Islamic rules in this respect were probably even less rigidly applied. Denham's narrative of the battle of Angala between Bornu and Bagirmi, for example, shows that the victorious Bornu took 480 horses and nearly 200 women from their adversaries and fellow Muslims.[57]

By no means all West African wars ended in a complete collapse of the enemy or the 'unconditional surrender' of one contestant to the other. There are numerous instances where diplomatic intervention and negotiation brought about a formal settlement, the most satisfactory of all outcomes whenever peace is regarded as a more favourable state than war. On the other hand war could sometimes lead to the virtual annihilation or absorption of one state by another or to the flight of whole communities – as was the fate of the Whydah after their defeat by the Dahomeans in the first part of the eighteenth century. Occasionally an undeclared peace of exhaustion might supervene, as happened in some of the Akan wars, especially those between the Ashanti and the coastal Fante, or among the Yoruba in the nineteenth century, but this rarely provided a lasting solution.

Instances of negotiated peace settlements, mainly those between the more centralized states of West Africa, were given in the last chapter. Among 'stateless' societies, diplomacy was naturally less professionalized but not necessarily less sophisticated. When the Kagoro grew tired of their intertribal wars – and they regarded fighting as on the whole a less strenuous occupation than work on their fields – those of their women who had relatives on the other side would act as intermediaries. Peace overtures took place under the protection of a broom-stick of plaited grass; the leading men held their palaver on the tribal boundaries, and concluded their agreement by killing and sharing a goat, and sometimes by exchanging hostages. Ibo peace negotiations were conducted usually through priests or the representatives of neutral towns, and under the auspices of national oracles, such as that of the Igwe-ke-Ala at Umunoha, and the Ihi spirit was invoked to guarantee an agreement. In such cases the outcome seems to have been a genuine adjustment to the changes brought about by the fortunes of war and an attempt to restore and stabilize political equilibrium. Perhaps the most surprising step towards this goal was the provision among the Ibo for money payments by each side to the other in compensation for those who had been killed, a warrior who had had three wives, for example, rating a higher payment than one with only two.[58] Thus the loser militarily might become the winner financially, a paradox not unknown in the aftermath of two world wars.

Instances such as these may serve to correct the view that the pre-colonial past of West Africa was a dreary tale of savage and almost continuous warfare, just as instances cited earlier may contradict the view that wars were conducted there with an almost ludicrous lack of spirit and conviction – contrary impressions which both stem from early writers such as Bosman and from that jaundiced nineteenth-century observer Sir Richard Burton. Similarly, the view, still adduced in some quarters, that these West African wars were actuated almost wholly by greed – or, to use a less emotive and moralizing term, by economic motives – may have been at least modified by the evidence which has been given as to the origins of the wars. For the most part, the wars of the West African peoples, and certainly those which took place between the more highly developed states, were discrete political events which arose from a complex variety of causes, by no means wholly or mainly economic, conducted with coherent ends in view, subject in varying degrees to mitigating rules of war, and generally bringing about, deliberately or incidentally, far-reaching change.

IV. Armies

Armies take many forms. In their human composition they range from citizen armies, the *levées en masse*, to compact groups of professionals, the first generally consisting of short-term soldiers, the latter constituting a long-serving élite. The principles of military leadership vary equally, from systems like feudalism where power is dispersed, to autocracies where it is concentrated. Again, the balance of arms varies, in the sense both of the weapons and of the men who wield them, reflecting the predominance either of artillery, or long-range firepower, or of infantry with their shock weapons, or of mobile or shock forces such as cavalry. Examples of all the main types of army are to be found in the military history of pre-colonial West Africa, reflecting the levels of political and technical attainment among the many peoples and states, their individual commitments and policies, and the varying physical conditions of the region.

There is evidence that nearly everywhere in West Africa all free adult males capable of bearing arms were liable for military service in time of war. Perhaps the earliest reference to mass enlistment for a particular campaign is the order by Mai Idris Aloma of Bornu, sometime in the late sixteenth century, that all his people (presumably only his adult male subjects are meant in the passage) should come to him armed with matchets, leather shields, and other weapons, and carrying their provisions; none, whether herdsmen or merchants or belonging to the Ulema, were to remain behind. The early authorities on the coastal peoples also mention this obligation of militia service. William Smith, however, writing of the customs of the Slave Coast in the eighteenth century, observed that after war had been decided upon, no one was pressed since 'the Manceroes being young run into it for the sake of plunder'.[1] In the nineteenth century, when firearms became widely available, this general mobilization (reminiscent of the English *fyrd*) was less resorted to as standing armies of various kinds were formed, but the obligation seems in theory, and sometimes in practice, to have been retained.[2]

The obligation of slaves to military service is unclear. The position seems to have been that a part of the household slaves of great men were always considered as having a special duty to accompany their masters to war and to fight with them, and this would apply *a fortiori* to royal slaves. Slaves played a large part in some of the early armies of the West Sudan. Whereas the cavalry of the savannah states was always, or nearly always, made up of freemen, a large proportion of the infantry was unfree; for example, in Mali the archers who formed the core of the army were slaves. In the armies of Bornu nearly all the soldiers were slaves, their commanders being themselves often slaves of the ruler.[3] Generally the citizen soldiers, but not the slaves, were required to provide their own arms, clothing and at least part of their food, but as weapons became more sophisticated and differentiated from those used in hunting, a greater proportion had to be made available from the arsenals of kings and chiefs. Many of the early musketeers were slaves whose weapons were provided by their masters.

For the most part, fighting was the concern of the younger men. Older men, however, often attained or retained positions as leaders or advisers to leaders, or were allotted tasks with the commissariat of the armies.[4] With the notable exception of the Dahomean 'Amazons', women seem not to have taken part in combat. Many of the armies were, however, accompanied into the field by women who performed such tasks as the preparation of food and 'medicines', bringing up ammunition and food to the front, and generally encouraging the warriors. But above all women were responsible for the care of the family farms and other property in the absence of their men.[5]

The widespread acceptance in West Africa of the principle that military service was incumbent on all freemen capable of bearing arms made it possible for the rulers of the larger states to mobilize a considerable number of combatants and supporters on appropriate occasions. European observers often made a special point of estimating the size of armies in the states they visited or of which they heard reports. In many cases their figures seem likely to have been exaggerated, both by the writers themselves, imbued with the traveller's desire to tell a good tale, and by their informants, eager for various reasons to create an impression of strength. Statements such as the often-repeated assertion, apparently first made by Bosman, that the ruler of the small state of Whydah in the seventeenth century could put an army of 200,000 into the field,[6] are clearly suspect. But even after making allowance for the tendency to over-estimate the numbers of combatants, it must still be concluded that the armies of West African states were of a surprisingly large size by European standards.[7]

Perhaps the earliest estimate of the size of a West African army is that by al-Bakri (who himself apparently never left Spain, relying on informants for his statements) about the empire of Ghana, where the ruler was able to 'put 200,000 men into the field'. More interesting than this figure, which is probably inflated, is the omission of any mention of cavalry, though al-Bakri does refer to the existence of horses at the capital – they were very small, and the king's horses were caparisoned in gold.[8] Accounts of all the later major states of the West and Central Sudan make it clear that

cavalry had become an essential part of their armies.
According to Leo Africanus, the ruler of Songhay (here
called the king of Timbuktu) 'hath alwaies three thousand
horsemen, and a great number of footmen', while the army
of Wangara amounted to 7,000 archers and 500 horsemen.
The Mai of Bornu was 'a most puissant prince ... Horsemen he hath in a continual readiness to the number of three
thousand, and a huge number of footmen'.[9] The Songhay
army which was defeated by the Moroccan invaders of 1591
was said to comprise 12,500 horsemen and 30,000 infantry.[10]
The infantry armies of the forest and the coast, on the other
hand seem to have been generally smaller. Bosman writes
of the Guinea coast that 'A National Offensive War may
well be managed here with four Thousand Men in the field,
but a Defensive requires more'. He gives no reasons for this
rather surprising claim, but goes on to say that while armies
rarely exceeded 2,000 in number, the Fante fielded 2,500
and the army of 'Aqamboe' was even larger.[11] Accounts of
the size of Dahomean armies vary considerably, as the
armies may well have done, but it seems that they did not
exceed 15,000 to 16,000 men and were often much smaller;
in 1774 the army was a mere 3,000.[12]

It was tentatively suggested in the last chapter that the
nineteenth century saw an increase in both the scale and the
severity of West African warfare. This is supported by evidence as to the size and composition of the armies, evidence
which becomes increasingly reliable as more and more
experienced observers recorded their findings. It has, for
example, been calculated on the basis of figures given by
Barth, that painstaking traveller, that the Sokoto Caliphate
could call on a total cavalry force of some 25,000 to 50,000,
with an infantry force of from five to ten times that of the
cavalry. This vast array would be made up of the Sokoto
army itself reinforced by contingents from the constituent
emirates; rarely, if ever, was it deployed as a whole. The

average size of the emirate army was about 10,000 men, but the ruler of the important state of Kano could call on some 7,000 horsemen and 20,000 or more foot. The Bornu army consisted of some 7,000 or more horsemen with a larger but indefinite number of infantry.[13] This apparent increase in the numbers taking the field is somewhat surprising since it took place at a time when the citizen army was being superseded by trained bodies of professional soldiers of whom a growing proportion was armed with firearms, a development which might have been expected to lead to a reduction in the size of armies.

In the forest states the position is rather less clear. For example, although a count in cowries was said to be kept of an Ashanti army, Bowdich and Dupuis differed widely about Ashanti military strength in the early nineteenth century, the former putting it at some 204,000, the latter at about 80,000, excluding camp followers.[14] Yet it is doubtful if an Ashanti army of even the lesser figure was ever put into the field. A more realistic estimate for the army of a forest kingdom at this time is that of 4,000, given in Dahomean tradition for the Oyo force defeated by king Gezo about 1820.[15] Better evidence is available from Yorubaland later in the century. The armies of Ibadan, Ijaye and Abeokuta which fought in the Ijaye War of 1860–5 were estimated by European and North American missionary observers at over 60,000, over 30,000 and about 20,000 respectively,[16] figures which, although possibly rather too high, bear out the belief that Yoruba armies were growing much larger in the second half of the century. Similarly, the widely-based coalitions of the Sixteen Years War led to a much greater scale of warfare. The Ibadan camp at Kiriji was estimated as containing about 60,000 souls and that of the Ekitiparapo as containing about 40,000. These figures include, however, the wives, children and attendants of the warriors so that, Akintoye suggests,

the fighting strengths should be scaled down to 30,000/
40,000 for the Ibadan and 20,000/27,000 for the Ekitiparapo
armies.[17] The Ijebu army which encountered the British on
the Yemoji in 1892 was reliably estimated by their professional adversaries at between 7,000 and 10,000 men. But
this evidence about the number of troops in the field is less
significant than that as to the composition and weaponry of
the Yoruba armies. These were coming increasingly to con-
sist not of militia but of the trained and seasoned retainers,
or 'war boys' (*omo ogun*), of the great chiefs, of whom
after about 1885 a large number were armed with single-shot
rifles, of much greater efficiency than muskets. Finally,
Samori's field armies, at the very end of the period under
review, varied between about 12,000 and 20,000, similarly
armed with rifles.[18]

The recruitment of women into the Dahomean army
seems to have been unparalleled elsewhere in West Africa,[19]
and the 'Amazons' (as they were dubbed) excited the interest
of the European writers and travellers of the eighteenth and
nineteenth centuries. The status of these female warriors was
that of royal wives, and as members of the King's household
they formed a part of his personal military force in battle
and of his bodyguard. Their organization in war, in two
wings, left and right, was similar to that of the male army.
The numbers of the force presumably varied, though perhaps not to the extent of those given by travellers; Dalzel
describes them as totalling not less than 3,000, of whom
several hundred were trained to arms. Burton reported that
two-thirds of them were celibate; that this was a requirement
of their situation appears from the trial of 150 Amazons
found to be pregnant, with their paramours. Elsewhere
Burton avers that 'a corps of [female] prostitutes' was kept
for the pleasure of these soldieresses. The origin of the force
is ascribed by Snelgrave to the reign of Agaja (1708–1740).
After being disastrously defeated by the Oyo and being faced

by an invasion of the 'old Whydah' (returning to recover their former land), Agaja is said to have reinforced his depleted army by ordering 'a great number of Women to be armed like Soldiers' and marched against the enemy. Thereafter the regiment seems to have formed an effective part of the kingdom's fighting force. Burton remarks on the physical superiority of the Amazons to their male counterparts, though he allows the men to be more intelligent. Amazons played a leading role in such fierce engagements as the assault on Abeokuta in 1851; on this occasion the Egba only discovered that women were among their opponents when they were preparing to castrate a prisoner, and in their ensuing fury redoubled their efforts to repel the attack.[20]

The relationship of enlisted warriors to the total population of West Africa is a matter about which little is known or can be surmised. Clearly, among the stateless peoples all, or nearly all, men capable of bearing arms were expected to take the field against an enemy, thus giving what Andrzejewski terms a high, or very high, 'Military Participation Ratio' (MPR).[21] This general obligation, deriving from ancient and more primitive times, underlay the militia armies of those who had attained statehood, and diminished as more sophisticated means of waging war were introduced. The only contemporary attempt to estimate the proportion of warriors to civilians in a West African state seems to have been Bowdich's assertion that the military forces of Ashanti amounted to about one-fifth of the total Ashanti population. This is considerably higher than figures for European armies of roughly comparable armament: the exceptionally large army raised by Edward III of England for his siege of Calais in 1346–7 probably amounted to only about one per cent of the English population. But in any case Bowdich's estimate cannot be reconciled with his statement in an earlier passage of his *Mission* that an Ashanti army was composed mainly of tributaries and allies under Ashanti officers.[22] Thus, on the evidence at present available, it does not seem possible or profitable to attempt to apply generalizations here such as those worked out by Andrzejewski about the causes and consequences of a high or low MPR.[23]

The character and organization of a West African army were strongly influenced by the physical features of the country in which it was recruited and mainly operated, which as indicated in the first chapter determined in particular the feasibility of the use of horses and similar animals. In the savannah of the West and Central Sudan cavalry constituted the most important arm, at least until the mid- and later nineteenth century when firearms became more prolific and efficient and the horseman and his mount correspondingly more vulnerable. The Mali, Songhay, Bornu, Habe (Hausa), Mossi, Futa Toro, and Dagomba armies are all examples in which cavalry played the leading role. Usually these cavalry forces formed an élite in which commoners were not allowed to serve. They were also very costly to raise and maintain. In the forest and coastal states horses were little known and little used, and cavalry did not exist. Among the ancient Yoruba kingdoms, whose territories lay mainly in the forest belt, only one possessed a cavalry arm; this was Oyo, for long the most powerful, whose capital and the greater part of whose territory were situated north of most of the other Yoruba and within the woodland savannah. According to both Bosman and Snelgrave, the Oyo armies of their times – the late seventeenth and the early eighteenth centuries – were composed entirely of cavalry.[24] Though this seems likely to have been true only of such long-range expeditions as those which made their way south-west from Oyo through the gap in the forests which leads to the coast east and west of the Volta estuary, there is little doubt that cavalry formed the most important part of the Oyo army and was the arm which enabled the Alafin to extend their kingdom and create an empire. Cavalry continued to be used by the main successor state to Oyo, Ibadan, which had been founded about 1830 on the forest edge. As late as 1881 Ibadan horsemen, left behind to guard the city as there was little use for them on the gun-dominated battlefield at Kiriji, charged and routed an army from Ijebu Ode, for whose troops horses were an un-accustomed and alarming sight.[25]

There is evidence that in the nineteenth century cavalry forces in the more highly developed armies of the savannah were differentiated into heavy and light troops, a develop-ment which can be traced (as will be seen in the following chapter) to five centuries earlier.[26] The differentiation was based on the type of horse used and on the operations which could be appropriately performed by horsemen with such mounts, and was reflected in the arms and armour of the warriors. The heavy cavalry, mounted on the larger and stronger imported horses (which can here be termed 'chargers'), were armed primarily with heavy thrusting spears (or lances), and usually also with swords.[27] Such a force could terrify a foe – especially an army of pagan footsoldiers – as they advanced under the cover of a barrage of arrows from accompanying archers. It could nevertheless be overthrown by determined infantry, as happened at Tab-kin Kwatto in 1804 when the jihadists of Usman dan Fodio, at that time without horses, defeated the mounted Gobirawa. Again, cavalry were naturally of little use in the siege operations against walled towns and camps in which, as will be seen, much of the warfare of West Africa con-sisted.

Light cavalry, riding the smaller, indigenous ponies and armed primarily with light throwing spears – javelins – were effective in patrols, skirmishes and flank actions; on occa-sion they also took part in frontal engagements, as did the

Shuwa at the battle of Musfeia in 1823.[27A] Though the carrying of bows and arrows by mounted warriors has been recorded, it is probable that this was for dismounted action only. Finally, there is some indication that towards the end of the nineteenth century cavalry were carrying pistols or carbines as secondary armament.[28]

In some armies, cavalrymen were accompanied by additional horses. These were led by dismounted squires whose duties were to keep such spare horses fresh in case of their being needed by their masters during the battle and then after a battle to mount and round up captives and booty. Alternatively, these horses were available for flight after a defeat, and Fisher remarks that to go into battle with one horse only or even on foot was an indication of a leader's resolve to fight to a finish.[29]

Cavalry may win a battle, but infantry are needed to win a war. Thus, in West Africa as elsewhere, infantry normally provided the backbone of the armies. Apart from their leaders and their immediate followers, who were often mounted, the armies of the forest and coastal states were entirely composed of footsoldiers, while from early times infantry also predominated in numbers in the armies of the West and Central Sudan, as al-Bakri's account of eleventh-century Ghana and Leo Africanus's account of Songhay show.[30] West African footsoldiers up to and in many cases beyond the seventeenth century, usually carried a shield and dagger and were armed, as their main weapons, with heavy, thrusting spears, a bundle of javelins and sometimes also swords. The famous Kanembu spearmen, for example, carried a large spear and three or four javelins, as well as their great shields. Until well into the nineteenth century they formed an important part of any Bornu army; Denham (himself an infantryman, who had fought in the last years of the Napoleonic wars), noted this and added that it was the infantry who usually decided a battle in Bornu as in

Europe. Thirty years later, however, their importance had dwindled, which Barth ascribes to their lack of zeal and loyalty for the new dynasty in Bornu.[31] The other important footsoldiers were the bowmen who, although to be classed as infantry, functioned as artillery support rather than in close fighting. They are recorded as forming a part of most West African armies, but their usefulness was naturally greatest in the savannah (though in the forest the shorter range but more powerful crossbow came into its own); in Ghana, according to al-Bakri, more than 40,000 of the 200,000 men whom the ruler could (allegedly) put into the field were archers. From the late sixteenth century the bowmen and spearmen were gradually supplemented and then at varying points in the nineteenth century largely displaced by musketeers.

Firearms were introduced into the Central Sudan as early as the reign of Mai Idris Aloma in sixteenth-century Bornu, when a number were obtained from Ottoman sources together with a contingent of trained musketeers. By the eighteenth century guns were becoming known in Hausaland, Nupe, Darfur, and Wadai,[32] and though in no part of this region was their use on a significant scale before the mid-nineteenth century (for reasons which will be discussed in the following chapter), a few trained musketeers often found a place among the infantry. On the Gold Coast firearms began to be imported at the end of the sixteenth century, mainly by English and Dutch traders. During the 1690s and the first quarter of the eighteenth century, the musketeers, as Kea writes, 'emerged as the principal military arm' in the armies of the kingdoms of the Gold Coast, and 'although archers still played a significant tactical role in battle ... the spearmen ceased to exist as a military factor'.[33] Among the Ashanti, access to the traders selling guns on the coast was an important issue from the early eighteenth century,[34] though guns 'remained the exception rather than the rule' in the land to their north.[35] But the impact of firearms was uneven, and their spread among the people living on and near the coast and in the forest belt was neither rapid nor revolutionary in its effects. In Yorubaland, for example, little use was made of guns until the second decade of the nineteenth century and it was not until just before mid-century that they became common.[36]

In addition to the land forces of West Africa, some attention must be given to the fleets of war canoes which were maintained on the lagoons of the south, on the great rivers, especially the Niger, on Lake Chad, and (apparently to a much less extent, though there is little evidence) on parts of the coast.[37] These canoes, almost always 'dug out' from the forest trees, were sometimes of great size, being capable of carrying on the calm waters of a lagoon about 100 fully equipped warriors, but the average war canoe was considerably smaller, and consequently considerably more stable and manoeuvrable. In many cases the canoes were used primarily as transports, but there are sufficient examples of battles fought from them on lagoons and in the Niger Delta to warrant the conclusion – *pace* Turney-High[38] – that a form of naval warfare was practised in West Africa.

The earliest reference to this warfare seems to be the account by the thirteenth century geographer, Ibn Sa'id, of the fleet which the ruler of Kanem maintained on Lake Chad and with which he made 'many raids ... on the heathen people surrounding the lake, capturing their vessels, killing them and taking them prisoner'.[39] Far up the Niger the empire of Songhay used its large fleet of war canoes for the rapid transport and provisioning of its mainly infantry army, and it has been pointed out that most of the great battles of Songhay history took place along the banks of the river.[40] There are also early references to naval warfare in the writings of the Portuguese and other visitors to the coast from the late fifteenth century onwards. It was observed that the Bissagos islanders harried their neighbours from the sea and ventured up the rivers of the mainland in their war canoes, while the men of Mina on the Gold Coast made 'frequent depredations by sea' on the coastal villages of the Komenda.[41] No account has been traced of any battle taking place at sea, but on the lagoons east of the Volta naval warfare was conducted among the Aja, Ewe, Egun, and southern Yoruba. Moreover, just as the canoe was an important means by which long-distance trade was conducted in West Africa, so also did it enable states to send expeditions against distant objectives, as, for example, in the extension of Songhay rule along the Niger and, more specifically, the raid carried out by a Lagos fleet against a French trading fort in the Benin river in the late eighteenth century.[42] The possession of canoes and knowledge of their handling sometimes gave significant advantages to a people when confronted with a more powerful enemy not so equipped. It enabled the Songhay to delay the advance of the Moroccan invaders of 1591 and the 'old Whydah' to survive the repeated attacks of the Dahomeans in the eighteenth century.[43]

It must now be asked what, if any, general characteristics can be discerned in the military organization of the pre-colonial West Africans. Success in battle depends more on leadership and organization (for a leader must have subordinates and a staff to be effective) than on weapons. A classification of West African military structures on such lines as those suggested by Andrzejewski cannot be undertaken here since detailed evidence from which to work out Andrzejewski's three criteria – the Military Participation Ratio, the degree of subordination to the leader, and the degree of cohesion in the army – is largely lacking.[44] Moreover, West African armies exhibited a bewildering variety of types, ranging from the village war clubs of the Kom and the age-grades of the Ibo[45] to the elaborate military hierarchies of the Hausa-Fulani, the Oyo and the Daho-means.

The position of the ruler in war provides a starting point for the inquiry. In the first place, just as the decision to make war was one often either shared by the ruler with or taken by a council, so the conduct of war itself was generally determined by the ruler in consultation with his war chiefs or even left wholly in the hands of these (usually hereditary) specialists. It was sometimes necessary to obtain the agreement of the troops themselves, even before an individual engagement, as did Idris Aloma before attacking Maija in the Habe kingdom of Kano. Secondly, the ruler himself seems only exceptionally to have been his nation's leader in battle. The Mossi kings (or Naba), for example, while probably originating as war leaders, by the seventeenth century were delegating their authority to field generals, and this seems to have been a common practice. Neither the Habe kings nor their successors, the Fulani emirs of Hausaland, normally appeared on the battlefield. At Zaria, even when the emir was himself present, the army was under the orders of the Madaki. The king of the Jukun remained at home during war, unless matters went badly for his army when his duty was to seize the spear of Kenjo, the national hero, and show himself to the troops. The kings of Dahomey occasionally went to war, and Gezo is said to have been killed in action in 1858, but in battle they took a subordinate place to the commander-in-chief, the Gau. Among the Oyo, the Alafin sometimes took the field until, after the death (or capture) at the battle of Ilorin (*c.* 1835) of Oluewu, it was decreed that he should remain in his palace. In Bornu, some campaigns were led by the Mai in person (as, for example, by Idris Aloma), others by the Wazir, and others again by the governors of the districts most concerned.[46]

It can be concluded that the operational control of armies

in West Africa generally devolved upon a non-royal commander whose title often stated or implied that he was the
leading war chief. Sometimes the commander occupied his
post by virtue of his leadership of one arm of the forces,
as, for example, in Kano and Maradi where the Madawaki
and the Kaura respectively were in command both of the
cavalry and of the army as a whole, or similarly in the
Mossi kingdoms where the commander, the Tapsoba of
Oula, was also master of the bowmen. In Oyo the metropolitan army was commanded by the Bashorun, the leading
member of the Oyo Mesi (a council which had both political
and military functions); the provincial forces were grouped
into two armies, under the Onikoyi and the Okere for the
'left' and 'right' hand provinces respectively. Superimposed
on this arrangement, probably in the seventeenth century,
was an overall commander, the Are-Ona-Kakamfo. The
Are, like the Kaura at Maradi, was required – presumably
as a political precaution – to live outside the capital. In
Dahomey the army was commanded by the two chief ministers of the king, the Migan and the Meu, but in the field
the commander was the Gau. The Benin army was commanded by the principal town chief, the Iyase, until in the
early eighteenth century, his place for this purpose began
to be taken by the Ezomo. In 'stateless' societies, arrangements for leadership in war were naturally looser and
simpler. The Kagoro, for example, had no war chiefs; when
war broke out they simply chose the man most skilled in
fighting as leader, while their chiefs followed behind to en-
courage or drive on the warriors.[47] Similarly, naval commanders were appointed, such as the Hi Koi, or 'chief of
the canoes' in the Songhay empire (with a civilian counterpart, the Hari Farma, 'chief of the waters'). In the Hausa-
speaking countries along the Niger, Barth reported the
presence of a Sarkin Ruwa, or 'king of the water', who
apparently combined military and civil roles, as did the

Maralegha in the Logone capital who forbade him to look at the river there.[48] All these offices seem to have been filled by men in whose families they were hereditary, but presumably choice was made among the candidates of those with military or naval experience and reputation.

It seems likely that most early chieftancies originated in one form or other of war leadership. But since all those societies of pre-colonial West Africa about which sufficient evidence is available exhibited such a rich and confusing variety of chieftancy institutions, few other generalizations are permissible. One broad distinction does appear; between those states in which apart from the civilian chiefs there was a class or were classes of chiefs who were concerned wholly or principally with war, and thus were capable of constituting a war council of professionals as well as leading soldiers into battle, and those where no specific war chieftancies seem to have existed and leadership in war devolved upon chiefs whose roles were in the first place political. The first type appears to be the more numerous. Examples abound both in the savannah and in the forest: Bornu, the Habe and Yoruba kindoms, Dahomey, all these exhibited this phenomenon, even though in some cases the high command itself was exercised by a chief whose office combined both military and political functions. The Mossi provided a variant of this first type since the only chiefs in their society seem to have been military ones, and these in peacetime assumed political roles. Similarly, the *asafo* companies of the Fante states were organized for both war and peacetime communal activities and their leaders had roles which reflected this duality.[49] In most of these examples, the war chiefs were assigned by their offices and title to the positions which they occupied, at least theoretically, in the army when it set out to war, described either in terms of the order of advance in column, as, for example, the van, the main body, and the rearguard, or of the left and right flanks of the army deployed into line, or a combination of these two.

Perhaps the best example of the second type is the king-dom of Benin where, according to a recent writer,[50] there were in the nineteenth century alternative commanders, the Ezomo, assisted by the Ologbere, and the Iyase, with the Edogun as his second-in-command. None of these derived their titles from their military functions, and there is no information as to how they selected the subordinate officers who in war presumably bore much of the responsibility for leadership and to some extent must have corresponded to the war chiefs of the first type. Again, Whydah, before its conquest by Dahomey, seems to have belonged to this second type since Bosman observed that command in war was entrusted there to 'an ordinary Person, whilst the Captains and Chief Men out of Fear stay at Home'.[51]

The problem next arises as to what were the battle posi-tions of the private soldiers – the ordinary combatants – especially the footsoldiers. Did they, in particular, serve in war under the command of those whom they also recognized as their chiefs in peacetime? In states of the first type described above, this was probably largely the case, although it is not clear what then happened to those whose superiors were civil rather than military chiefs (who themselves did not appear on the battlefield except, perhaps, as attendants on their sovereign or his representative). In the intermediate case of the emirate armies of the nineteenth century Sokoto Caliphate, detachments raised by a district chief usually served under that chief's captain, while the chief himself remained with the army commander. In the second type, the problem seems easier since the whole warrior population, except for those owing direct allegiance to the crown or to the field commander, could be directed (usually with members of their age-grades) where they were most needed. But in the absence of evidence about early conditions, this remains speculative, and whatever systems originally obtained have been obscured by the subsequent development, in many cases only during the nineteenth century, of standing armies in which a man's post in battle, as well as his place in camp and on the march, was determined by professional superiors. Before the rise of the latter is described, however, it is proposed to examine briefly the allegedly 'feudal' nature of a part of West African society in so far as it applies to the recruitment of the population for war and its deployment in battle.

There has been considerable debate as to whether feudalism existed in Africa or, more precisely, whether the term 'feudal' can appropriately be applied to any part of African society.[52] It is fortunately unnecessary to enter here upon the details of the debate, and it may at once be granted that if feudalism is equated with either a certain level of agrarian technology (as does Goody), or chivalry, then the use of the word and its derivatives to describe any society in pre-colonial West Africa is improper. Nevertheless, it is still useful to ask whether feudalism existed in West Africa in the broad sense of the granting or holding of land in return for performance to the lord by the land- (or 'fief-') holder and his dependants of military service. In this sense, the answer must be that it did exist, although the obligations were far less clearly defined than in European feudalism. The nearest approximations to the latter in West Africa were in the savannah states. This was mainly because in these states cavalry was an important, often the most important, part of their armed forces and the mounted warrior, as in European feudalism, was the characteristic element in the military system. The burden of providing and above all maintaining the costly equipment needed for the cavalryman – the charger, the protective armour, and weapons – could not be met either by the rank and file or wholly by the king or overall commander, but was divided among the great men of the kingdom roughly according to the territorial holdings whose revenue made it possible for them to assume the obligation. (The cases where all cavalry horses were owned and stabled by the central power seem to be exceptional, though two are cited by Law.[52A]) Thus, in West Africa as in Western Europe predominance of cavalry warfare was among the causes of political decentralization, and those who have described the Habe, Nupe and Hausa-Fulani military organizations[53] have not hesitated to use feudal terms in doing so. For example, the adoption of cavalry by the armies of the Fulani jihad soon after their victory over the partly cavalry armies of Gobir at Tabkin Kwatto in 1804 led to the development among the Muslims of a warrior aristocracy, apparently resembling that which had existed among their Habe adversaries, and the *hakimai*, or district chiefs who were required to provide a fixed number of trained warriors for the royal army, and also maintain the walls of their towns, are often referred to as 'fief-holders'. But feudal terminology has also been used, and with some appropriateness, in describing the society of the forest kingdom of Benin where there was no cavalry and only the great chiefs were mounted, and yet 'the Oba's domains were divided ... into a large number of tribute units'. The 'fief-holders' were mostly the leading nobles of the capital (*Eghaebo*) – for whom 'tenants-in-chief' would be a more suitable term – and below them came the village heads (*Enigie*) and elders (*Edionwere*), paying tribute and rendering service as was directed from the capital.[54]

But though the cost of a cavalry force was exceptionally high, other forms of warfare could also make demands beyond the means either of a central government or its ordinary citizens. Boats resembled horses in this, and two nineteenth-century travellers, though writing without any pretensions to making a technically close analogy, describe the raising of the fleet of war canoes on the Lower Niger in feudal terms. When proposing to engage in war the king of Abo sent to his tributary chiefs who then provided him with a number of war canoes according to the sizes of their villages – 'some four, some only one canoe each', while some of the chiefs in the capital sent as many as six. 'Thus the villages appear to be held as military fiefs', wrote Allen and Thomson, adding 'but we have no opportunity of noting further the resemblance to the feudal system'.[55] As has already been admitted, such usage is justified only when 'feudalism' is understood in the broadest sense of revenue-yielding land-holding (or even, community leadership) which is dependent upon the rendering of specific military service, but even when all other analogies with the classic feudalism of Europe are eschewed, it throws light on an important and fundamental link between societies at war and at peace.

Although most West African armies up to about the middle of the nineteenth century were basically militia, regular or standing forces did exist. In the Mali empire, slaves trained as archers formed the core of the army and were apparently full-time, long-serving soldiers. In sixteenth-century Bornu a similar group of specialists, the musketeers, formed a trained standing force, partly composed of foreign mercenaries and partly of household slaves but of great tactical importance, as was shown in the battle of Amsaka, described by Ibn Fartua.[56] Curiously, there is no evidence of the use of firearms in Bornu after the seventeenth century[57] until the mid-nineteenth century, but the Kanembu spearmen noted by Denham and later by Barth formed a similar specialized, long-service corps in the Bornu army, which consisted otherwise of the households of the Mai and his great courtiers, numerous in slaves and clients. In Oyo there was a body of 70 junior war chiefs, the Esho, all nominated by the members of the Oyo Mesi council and confirmed by the Alafin. These constituted, according to Law, a 'semi-standing army of specialist soldiers,'[58] their specialization residing in their ownership of and ability to manage and maintain a war horse.

The evidence suggests that it was the introduction of new weapons and forms of warfare demanding special skills and regular training which led to the development of standing armies. A yet more obvious reason for this, or at least the maintenance of a professional force once it had been formed, was the need to make continuous provision for internal and external security. This second requirement was met in many cases by the existence of a permanent royal bodyguard which then formed the nucleus of a national army, even though in some cases (like that of the Mossi[59]) it might itself be debarred from taking part in a foreign war. The Dahomean armies of the eighteenth and nineteenth centuries consisted partially of regular troops who were dependants of the king and normally lodged in his palaces, while the war chiefs (*Ahwangan*) were themselves followed to war by their dependants[60] who, like those of the king, may be assumed to have had some regular military training and status. This seems to have been the pattern in many, perhaps most West African states, and it was probably arrangements of this kind which led the Sieur d'Elbée to write of the troops of Allada in the seventeenth century that they 'are not Militia, like those of *Whydah*, but regular Troops, constantly kept-up, who only want Fire-Arms and good Officers, to be able to recover the Provinces...'[61]

It was suggested in the last chapter that the nineteenth century was a time of deepening political trouble in West Africa when warfare became increasingly prolonged and bitter. The evidence is scattered and far from conclusive, but it is clear that the century did see radical changes in the conduct of war and in military organization. To a large extent these changes were due to an increasing use of fire-arms and to the increasing efficiency of those arms. Smaldone has produced considerable evidence to support his

claim that the general adoption of firearms in the Central
Sudan during the latter half of the nineteenth century had
no profound impact on the military and political situation
but did nevertheless enhance the importance of the infantry
at the expense of the cavalry and led to the formation of
large-scale standing armies, mainly composed of slaves and
slave-officers and closely controlled by the king who thus
increased his power to the detriment of that of his great
feudatories.[62] Thus in the Sokoto Caliphate, Bornu, Nupe
and elsewhere in the Central Sudan, far-reaching military
developments were set on foot in the 1860s, to be brought
to a premature end by the imposition of colonial rule
towards the end of the century.

Somewhat similar changes were taking place in the forests
far to the south, where the long drawn-out wars of the
Yoruba were fought by armies of an increasingly professional kind, made up of 'war boys' who were soldiers first
and farmers if at all only second, and were led by 'new
men' who had few connexions with the chiefly classes. These
developments are well illustrated by the rise of Ibadan from
a war camp founded about 1830 to be the centre of an
'empire' which largely replaced the old Oyo kingdom and
was then challenged in the Sixteen Years War of 1877 to
1893 by the coalition known as the Ekitiparapo. In
'republican' Ibadan a new set of chieftancy titles was
created, divided into civil and military orders, while in many
other Yoruba towns and kingdoms the old war chieftancies
were relegated to the background and new titles and offices
created to meet the needs of the time. Many of the new titles
were copied from Oyo, the title of lesser war chiefs at Oyo,
'Balogun', being especially popular, and even that of 'Are'.[63]
But in contrast to what was happening in the Central Sudan,
the changes in warfare among the Yoruba in the nineteenth
century seem to be connected less with the increasing use
and efficiency of firearms, than with the political and econ-

omic factors mentioned in the last chapter and, in the case of Ibadan at least, with the increasing numbers of slaves made available by successful wars for work on the farms, thus enabling ever more professional armies to be maintained.

West African armies were often reinforced by allies and by mercenaries. Combinations against a common foe, forged by diplomacy, were a familiar feature of the political scene, as has been seen in the preceding chapters. Sometimes reasons of state could produce surprising alliances, as in the adherence of the Muslim power of Ilorin to the anti-Ibadan coalition of 1877. Tributaries were also called upon by their overlords for active co-operation in war. As mentioned above, the Ashanti army, or so Bowdich alleged, was sometimes composed entirely of tributary contingents, while the Oyo on occasions in the eighteenth century required the Dahomeans and the Mahi to make war on their behalf. Dalzel writes of the attack on Badagry in 1784 that 'The operations of the Dahomean army were directed by the Eyeo (Oyo) messengers, who had conducted them hither; and nothing of importance was undertaken without their concurrence'.[64] On the Gold Coast mercenaries were readily available for war service. An injured nation might hire another's troops 'for less than two thousand Pounds sterling', Bosman says, adding that the Dutch trading company had prosecuted a four-year war for under £6,000, for which sum they had successively hired five different sets of mercenaries.[65] The Whydah usually relied on mercenaries for their defence, despite their own large reserves of manpower, and the Aro in the hinterland of the Niger Delta were even more notoriously dependent on the paid troops provided them by their neighbours.[66] On the lagoons between Lagos and the Volta, local rulers called often on the services of the Iso canoemen, renowned for their skill in naval warfare.[67] In the Sudan mercenaries were equally available and valued. By

1712 the army of the Ruma – the descendants of the Moroccan invaders of 1591 – had come to be largely composed of Arab, Berber, Fulani, and Tuareg mercenaries, while in the next century the Bornu were making much use of the restless Shuwa Arabs as light cavalry, though these were admittedly a 'band of robbers'.[68] Nor did the leaders of the Fulani jihad in Northern Nigeria scorn the participation of mercenaries in their holy enterprise, as Usman dan Fodio's handbook reveals.[69] On occasions Muslims fought not merely as allies of but in the armies of pagan states; there were said to be 7,000 Muslims in the Ashanti army which invaded Gaman in the eighteenth century, though the figure is probably exaggerated, while Meredith reports the presence of a Hausa officer from Katsina and his men with the Ashanti who besieged the British fort at Anomaba on the Gold Coast in 1807.[70] Europeans also sometimes sought aid from the West Africans in their wars in the region, while conversely the Africans asked for European help in their wars, a topic to which further reference is made in chapter VII below.

Foreign observers, with preconceptions about military dress and the externals of discipline, tended to ridicule the appearance of West African armies, especially the infantry armies of the forest. Barbot wrote that the Benin soldiers on the march went 'skipping and dancing into measure and merrily, and yet keep their ranks, being in this particular better disciplined than any other *Guinea* nation.'[71] Yet apparent disorder was by no means real disorder. Just as the chiefs knew their traditional places on the approach march and in battle, so in a mêlée could friend be distinguished from foe by dress and accoutrements and also by those facial and body sacrifications and painting which were prevalent among West Africans. Dapper describes the warriors of the Gold Coast as painting themselves with lines, crosses and serpents in white, yellow and red, and wearing a horned hat of crocodile or leopard skin adorned with red feathers.[72] Elsewhere dress was less picturesque. The Egba soldier of the mid-nineteenth century wore 'a sort of "trews" and a war jacket', the latter sewn with charms and cowrie shells, and this seems to have been the uniform of Yoruba armies.[73] In the Central Sudan, despite widespread assimilation of Muslim culture, striking contrasts in dress were observed. The Kanembu spearmen went into battle wearing only a belt of skin and a few strips of cloth twisted round their heads, while the soldiers of the Sokoto Caliphate dressed generally in long shirts, usually dyed blue, and each carried a blanket slung over the left shoulder.[74]

Finally, an attempt to identify the general characteristics of the armies of West Africa requires that some attention be given to the problems of the commissariat. A limiting factor in all wars, and especially those which are mobile, is the supply of necessary provisions, above all of food and munitions, including forage for cavalry mounts and beasts of burden. In West Africa this problem was greatly simplified by the general requirement that troops should provide and carry with them their own food; naturally they also carried a quantity of ammunition (arrows, powder and shot), this being often disbursed by the king or by his chiefs. When private food supplies were exhausted, a West African army was expected to live so far as possible off the land, either by foraging for itself or by compulsory contributions levied from the surrounding countryside. Ibn Fartua pointed out the advantage of mounting a campaign in Kanem at a time when the dates were ripe. Four hundred years later Barth noticed that the villages in the area of the Bornu army were each required to send in two oxloads of grain for the use of the troops; when this source failed he saw the soldiers cutting the standing crops in enemy fields and threshing grain. Like Denham before him, he was critical of the inability or unwillingness of the Bornu command to ensure the supply of provisions, but this was no new phenomenon: armies operating in the savannah seem often to have gone hungry and to have abandoned sieges because of this. Water supply was a special difficulty in the dry season when most campaigns took place, and water bottles or canteens of plaited grass had to be added to the soldier's load. Barth mentions also the provision of fodder for the cavalry chargers and the asses in the supply train of these savannah armies. In Oyo such matters of administration were the responsibility of the 'master of the horse', Ona Olokun Esin, though it is not clear if this was an officer of the field army or an official of the Alafin's stable.[75] The savannah armies were well provided with beasts of burden, making use of oxen, mules, donkeys, and occasionally camels. In the forest states, where such animals were not found, the soldiers sometimes carried their mats and bedding on their heads, as well as powder and shot and food for a week or two, and Burton describes the Amazons as doing this. It was also usual for an army to be accompanied, as were the Egba when going to the Ijaye War, by what Burton calls 'a mob of sutlers' acting as head porters; in the previous century Snelgrave had observed that a Dahomean army of about 3,000 troops was followed by 'a Rabble of ten thousand at least, who carried Baggage, Provisions, dead Peoples Heads, etc.'[76]

For troops in the field the most pressing need was for a food which was compact, portable and durable, and did not require cooking. It was perhaps this practical aspect rather than a fear of effeminacy which led the Ibibio to forbid their warriors to eat 'soft chop' in wartime.[77] Each member of the armies of the Sokoto Caliphate was required to carry three weeks rations with him, consisting generally of dried cooked meats.[78] The Yoruba soldiers subsisted on parched beans and a special kind of hard bread made of beans and maize flour; the Dahomeans ate toasted grains and bean cake; the Ashanti on campaign mixed a little meal with the first water they came across, while their observations scouts, who spent days hidden in high trees, lived on pounded grain and 'war nuts'. Similar iron rations – parched corn pounded and mixed with water – was the lot of the common soldier of Bornu on active service.[79] The grain used in early times was probably mostly guinea corn, but in the nineteenth century, maize, which had been established on the coast by the seventeenth century, began to be found most suitable for soldiers' rations, especially as it yielded two crops yearly with little trouble.[80]

No regular payment seems to have been made to the soldiers of any West African army but, as was indicated in the last chapter, all could hope for reward from the various types of booty which a successful campaign afforded, especially slaves and the profit from their sale. If rewards were not forthcoming, morale would suffer and protests might be made. Barth describes one such occasion, when the soldiers of Bornu, disappointed of pay or plunder, paraded before the tents of their commanders, shaking and beating their shields; the Shuwa and the Kanembu, who usually went ahead of the army but had on this occasion been prevented from doing so, were especially angry.[81] Sometimes the dissatisfied deserted. Such desertions were frequent at the beginning of the rains when the citizen soldier expected to be released in time to plant his farm.[82]

Information of the kind given above about the different aspects of a people in arms may appear merely miscellaneous, and a summing-up can only be made in the most general terms. Clearly, among the many forces which make for change in society, war and preparation for war are among the most important. This much appears from any general historical survey, such as the first volume of the *History of West Africa* edited by Ajayi and Crowder, in which the major emphasis is placed on 'the theme of state formation', and the pages of which are studded with wars to the point of wearisomeness. Social scientists have lately come to see in 'conflict' a major source of social change. At the risk of being still more banal, it may be worth elaborating the idea a little further. Thus, it is not only war which brings change – integrating peoples, moving boundaries, forwarding technology, and so on – but the organizational effort needed to form and keep an army in being, whether in the field or on peacetime duties. Secondly, these changes are multifarious and by no means uniform. War may dissolve a society, it may strengthen it, and either result may come about in many different ways, and not just as the aftermath of victory or defeat. The extent to which war, or military life, is a 'slayer' of consanguinous societies, in Turney-High's phrase, is limited by the degree of participation – Andrzejewski's 'Military Participation Ratio'. If an army constitutes an élite, it is more likely to be a divisive than a cohesive force in the state. But such armies were not characteristic of West Africa and it was apparently only in the last years of the pre-colonial period that specialization developed to the degree which could produce an army of this type, a process which was interrupted by the conquerors and colonizers from Europe.

V. Arms and armour

Wars are probably won more often by superior organization and morale than by superior weapons. Turney-High writes, with some exaggeration, that 'weapons are merely tools used to facilitate' the practice of war.[1] Yet these 'mere tools' are of first importance in forwarding success in war, for the result of a battle or a campaign can turn upon them, and often too they hold the balance between peace and war. Moreover, the weapons and accoutrement of the warrior reflect the technological achievements of a society and are an index to its art and culture.

In West Africa, the range of basic weapons which was used in pre-colonial warfare was as wide as that used by West European armies before the onset of the nineteenth century, though in many cases the development of particular weapons had proceeded less far – which was not always a disadvantage since West Africa avoided, for example, such follies as the increasingly long and heavy swords of medieval Europe. The projectile weapons used were the bow, cross-bow, javelin or throwing spear, dart, and sling, and then the gun; the use of the blow-gun was apparently unknown.[1A] Among shock weapons, those decisive instruments in war, were the sword, club, lance, dagger, and fighting bracelet. Mobility was provided by the horse and other animals and by the canoe, while the shield and metal and padded clothing provided defensive armour.

The presence or absence of certain of these weapons in particular communities may make some basis for the kinds of generalizations which characterize Goody's work on *Technology, Traditions and the State in Africa*: 'The bow and arrow is essentially a democratic weapon'; 'With technologies of the bow and stone-tipped arrow, any kind of centralization is impossible. But with the introduction of metal, kingdoms are on the cards.'[2] To the present writer, however, the widespread use in West Africa, in varying degrees of sophistication, of all these types of weapons and their adaptation to the physical environment seems more remarkable than any postulated relationship to forms of society.

Evidence about these weapons is relatively plentiful. In the first place, much armament has survived from the wars of the last century, preserving the forms of earlier times, and some examples may be even older. A second source is the art of West Africa, in which sculpture plays a prominent part and the warrior attired for war has long been a favourite subject. Finally, European visitors to Africa from the fifteenth century onwards took a surprising amount of interest in the weapons which they saw in use, preserved against use, or being made, and accounts such as those of Dapper in the seventeenth and Denham and Barth in the nineteenth centuries include detailed descriptions and sometimes also drawings of these weapons.

The majority of the weapons were locally made. West Africa had an abundance of woods which were hard, pliant and workable, and of craftsmen skilled in carving and carpentry. Bowyers, shield-makers and similar artificers practised their crafts usually in family compounds which were under the special protection of town chiefs, and they had their own gods and rituals. Iron ore, mostly of the low grade known as laterite, is also widely distributed in West Africa, and the technique of working iron was known there from the early centuries of this millennium; moreover, by 1600 large quantities of iron were being brought to the coast by

European traders. Blacksmiths were numerous, usually working in closely-organized kin-guilds. As Bosman wrote of the inhabitants of the Gold Coast, 'with their sorry Tools they can make all sorts of War-arms that they want, Guns only excepted'.[3] A large part of this production must have consisted of lance-heads, iron points for arrows, and swords of various patterns – though, as a Yoruba chant implies, locally-made blades were unpopular as they were liable to 'twist and spoil'.[4] On occasion the blacksmiths were required to work in close co-operation with the army, as happened at the 'pagan' town of Amsaka during its siege by the Bornu in the late sixteenth century, when the Muslims could hear the anvils inside the town being beaten in a frantic effort to replace the defenders' exhausted stock of arrows.[5] The repair of guns was also undertaken,[6] but, unlike some of the Asian peoples, West Africans (with probably only one exception, at the end of the nineteenth century) did not reach the level of skill in iron-working needed for the successful manufacture of firearms,[7] and from the seventeenth century onwards these were an important article of import. In the seventeenth and eighteenth centuries cutlasses also featured prominently in lists of articles suitable for trade to the West Coast and in shipping lists, while sword blades continued to be imported until the end of the pre-colonial period. Between the thirteenth and sixteenth centuries, Tlemcen was the main entrepôt for the trade in blades, which came mostly from Marseilles, Bordeaux and Genoa; in the nineteenth century Solingen was the main source of supply. The setting of these blades was a very profitable industry in Kano.[8]

Before the individual weapons used by the armies of West Africa and those means of locomotion which by extension fall under this heading are considered, the association between the provision of weapons and the distribution of political power must be noted. When weapons of war are relatively simple and can be obtained by the warrior at low cost from local artificers or even made at home for his own use and ownership, an army, composed usually of militia, can be quickly and cheaply raised; such an army, however, is not readily susceptible to central control. That many West African weapons were of this simple type seems at least in part to account for the instability of the armies described by early European observers on the coast, and especially of their inability to sustain a campaign in adverse conditions or beyond the end of the dry season. This weakness was to some extent countered in the more centralized states by the storage of weapons and other munitions – poisoned arrows, for example – in royal arsenals. The demand for better-quality and more complex weapons, such as swords with steel blades and above all firearms, and other equipment and means of war, especially horses, had other results. These were expensive items, usually imported from great distances, which were beyond the means of the ordinary soldier. In some cases – and particularly perhaps, as Goody has suggested,[9] where firearms were concerned – this contributed to the power of the central authority since the king not only owned the arms but was also able in time of peace to retain them under his control in his palace. In other cases – and particularly those of cavalry horses and armour – the expense of providing war material might be met by the great men of the kingdom, at least in part, so that a decentralized government of a feudal type, as discussed in the last chapter, tended to emerge.

Swords

For many centuries the primary shock weapon nearly everywhere has been the sword, pre-eminent in hand-to-hand fighting and used both for stabbing an enemy and, less effectively, for cutting and slashing him. Goody maintains that in tropical Africa 'the sword was never of great importance (except ritually); the basic infantry weapon was the spear or javelin, with the shield for defence'.[10] The second part of this statement is well supported by evidence from the savannah in the pre-colonial period. Yet the undoubted importance of the sword in ritual and the large number of surviving swords, of both operational and ceremonial forms, attest to its having played a leading role in warfare over much of West Africa. According to Kirk-Greene, tradition in Bornu tells that the first dynasty of rulers there, the Sef, fought with swords, as opposed to their successors, the Magumi, who were distinguished by their spears.[10A] There are, moreover, references in travellers' accounts to swords being carried by cavalry, by archers and by ordinary infantry. In the cases of the first two, it was probably regarded as secondary armament, but for at least some of the forest infantry (for example, the Yoruba) it was a major weapon until the large-scale introduction of guns.

The pattern of two-edged straight or gently tapering sword, characteristic of the Middle Ages in Europe and North Africa, was common in the savannah and is found also in the forest. At Katsina in 1851 Barth saw a review of the Emir's army in which the cavalrymen wore such swords, and the state swords at Daura (the 'Sword of Bayajidda') and Argungu (the 'Sword of Kanta') are both of this type; at Katsina the state sword (*Gajere*, 'the short one') has a slightly curved blade, only one-third of its back edge being sharp.[11] According to Captain Jones, the swords used by the Egba soldiers at Ijaye were 'always straight, double-edged, about 3 feet long',[12] while the 'Sword of Ajaka' at Owo in Eastern Yorubaland has a slightly tapering two-edged blade some 4 feet in length.[13] A variant of the straight sword is the single-edged, narrow-bladed back-sword. Only one West African example of this is at present known to the writer,[14] this being almost certainly entirely of local manufacture, the decoration of the hilt relating it to either the Egba or the Ijebu of Yorubaland. Most surviving examples of straight swords have elaborate hilts, with quillons curving downwards (or, in the case of the backsword referred to above, upwards) as a *couvre-main*; those of Islamic provenance often carry inscriptions on the blade. As on European swords, the blades are sometimes indented with fullers – vertical grooves, picturesquely explained as gutters for blood but which were probably intended to decrease the weight.

Much more common in the forest lands than these swords is a type in which the blade slopes outwards from the hilt to reach its greatest width, often about two inches, just above the point, making an elongated leaf-shape, resembling the Hallstatt swords of the European Iron Age; others of this type are of uniform width throughout and culminate in a curved spade-like end.[15] It seems likely that it was swords of this latter kind which Dapper saw on the Gold Coast and described as 'a sort of Chopping Knives'. Such swords were strong, he adds, but extremely blunt.[16]

Apart from swords, the warriors of West Africa carried a variety of similar but shorter weapons for stabbing, including European-style daggers. There are surviving examples with both locally-made and imported blades.

Staff weapons

There are grounds for concluding that the spear in its various forms was the major weapon of both cavalry and infantry in West Africa. Where the sword was used for close fighting, the spear served to engage an enemy at arm's length by thrusting or at a much greater distance when thrown. Like the sword, it was a symbol of honour and office as well as a weapon of war, being often carried into battle as a standard, while spears handed down from ancestors, or symbolic representations of these, more decorative than useful, form part of the regalia of kings and chiefs.

Goody claims that there was no authentic development of heavy cavalry in tropical Africa since 'the heavy lance never replaced the javelin, and the coats of kapok, mail and Koranic leather charms worn by the horsemen were mainly a protection against infantry rather than opposing cavalry'.[17] Unfortunately no accounts seem to have survived of any important cavalry engagements in pre-colonial West Africa; the jihad of Usman dan Fodio presumably witnessed such action, as, for example, in the Alwassa-Gwandu campaign, but more was accomplished in local Fulani-led risings, and there was no single great clash between the armies of Sokoto and Bornu. Yet (to return to Goody's contention), the lance was an integral and in some cases the major part of the weaponry of West African cavalry. It seems normally to have been wielded at arm's length rather than carried at rest, despite the use of stirrups, so that the mounted shock combat evolved in medieval Europe, in which horse, rider and weapon charged as one, was not practised.[18] Nevertheless, the broad-bladed six or seven foot long lances of the Sokoto and Jukun cavalry and the double-headed lance of the Bagirmi, which could penetrate chain mail, qualify as 'heavy' equipment even by European standards and certainly by comparison with the light equipment of other of their fellow cavalrymen. Again, cavalry in the Hausa-Fulani armies increased the weight of the downward thrust of their lances by wearing two or three stone bangles, each of a pound or so in weight, on their spear arms.[19]

The fundamental distinction in spears is not by weight; it is between those intended for thrusting – 'lances' when carried by cavalry, 'spears' when used as an infantry weapon – and those intended for throwing either by infantry or horsemen, alternatively called 'javelins'. Throwing spears were naturally lighter and more centrally balanced than thrusting spears, and the distinction between the two is generally made in the vocabularies of those West Africans who used the weapon in these two ways. Spearmen on foot usually carried a bundle of several spears to provide spares. The problem of replacing a spear thrown by a horseman was more difficult. In some cases footsoldiers attended the cavalry with spare spears, as is shown in Denham's drawing of the battle of Musfeia. But the cavalry of the Central Sudan – again as appears from a drawing by Denham – also carried spare spears on their horses, the points fitting into a case suspended from the saddle. Footsoldiers coming up in the rear retrieved the thrown spears. In addition, West Africans, 'like Irishmens' as Jobson writes, threw darts, a smaller version of the javelin (or simply arrows used in close combat), barbed and 'full of crueltie to the receiver'.[20]

West African spears consisted of wooden shafts to which an iron head was socketed or (less often) tanged. These iron heads varied considerably, having broad, leaf-shaped, tri-angular, or more eccentric points, and in most cases carried barbs or reversed points. As with arrows, the heads were often dipped in poison before being taken into battle. Sometimes iron points were also socketed to the butts of the spears, providing the user with a backward stroke against an enemy coming from the rear and enabling the weapon to be stuck in the ground when out of use or to give a mount for a cavalryman springing onto his charger.

By the mid-nineteenth century spears seem to have largely fallen out of use in the forest armies and were replaced by muskets; Captain Jones, for example, saw 'but very few spearmen' at Ijaye in 1861.[21] In the savannah, spears re-mained a part of the cavalryman's equipment to the end of the century, and some infantry there continued to carry them. Denham describes the commander of the Sheikh of Bornu's expedition to Mandara in 1823, Barca Gana, as himself throwing (presumably when mounted) eight spears 'which all told' at 30–35 yards. But by mid-century Barth was inclined to discount the effectiveness of the Kanembu spearmen in the Bornu army.[22] Finally, a photograph showing a Fulani charge against the British invaders of Hausaland in 1903[22A] suggests that these heavy cavalrymen were using their spears as javelins rather than as lances, a development probably due to their despair of being able to close with an enemy armed with repeater rifles and machine guns.

Missiles

Originating in prehistoric times, the bow and arrow is held to be man's first invention for accumulating and storing energy. Whether it was independently invented in West Africa or reached the region by diffusion – perhaps from the Nile Valley – has not been determined but, as the rock paintings of the Sahara show, it was used in very early times by the possibly negroid peoples there. From the times of the first accounts by travellers and until well into the nineteenth century, archery (or, to use an ancient and equally applicable term, artillery) was an important item in nearly all West African armies. It provided long-range firepower and was especially valuable in open country, for example, in the savannah at the dry season when the grass was short. Bows may indeed still be seen in use by the hunter in the bush.[23] In warfare bows and other missiles were mainly infantry weapons, as in Europe and the Middle East, but some cavalry – for example, that of the Oyo and of the nineteenth-century Adamawa – seem to have used bows. Barth mentions seeing (to his surprise) a Bornu archer on horseback, and both Benin and Yoruba sculpture show mounted archers.[24] According to Bosman, bows were not much in use on the Gold Coast in the seventeenth century except among the dexterous Aquambo, but they feature among the descriptions of weapons there by writers from Jobson onwards.[25] Lander describes the Oyo in the first part of the nineteenth century as still engaging in frequent archery practice and as being so skilled that they could send their arrows through a small opening in a wall 'upwards of a hundred yards' distant, while Fulani (and presumably also Hausa) bowmen of the Sokoto Caliphate were equally accurate. But such accuracy was probably unusual and in any case, as Muffett observes, 'was not an essential criterion of effectiveness, especially against cavalry' – a horse, even when galloping, presents a large target. In battle archers were more usefully employed in firing barrages, some being capable of shooting two arrows at a time.[26]

The self or simple bow consists essentially of a bent stave of pliant wood and a bowstring of a sufficient elasticity. All known West African types may be characterized as shortbows, being from about two and a half to five feet in height (the English longbow was about six feet, or the height of the archer).[26A] They were cut from various types of hardwood,[27] while the bowstring consisted of twisted raw hide (antelope hide being a favourite), usually threaded through a hole at one end of the stave and then bound and tied frontally at the other end. Some extant Yoruba bows are fitted with handgrips of leather in the middle of the stave and have 'medicine' – miscellaneous substances believed to possess magical powers – enclosed in wrappings bound to the stave. The force of such bows at full draw varies between about 40 and 50 lbs, as compared with the 100 lbs or more of the English longbow. Their effective range was some 50 to 75 yards and at close range they were capable of killing an elephant. Bowmen often protected themselves against the bowstring by wearing an iron finger- or thumb-guard on their right hand and a leather bracelet on the left arm.[28]

The cross bow, a specialized form of bow more suitable for defence than for offence because of its slower rate of fire, was apparently in use among only a few of the forest peoples and seems to have been unknown in the savannah. No descriptions of this weapon have been found in the accounts of West African armament given by the early European and North African travellers, but a missionary report of a military review at Ijaye in 1861 refers to the carrying of 'great crossbows' by some of the troops (others had swords and shields, others bows and arrows, but most were armed with muskets).[29] A Yoruba proverb recorded by Bishop Crowther affirms that: 'A crossbow is not enough to go to war with; whom then do you dare to face with only a stick?' Crossbows have been collected from or observed in Yoruba country, Benin, Agbor (in Western Iboland), the Cameroons, and on the Gabon and Ogowe rivers.[30] All had stocks and staves of wood and were of roughly the same archaic type, the stock being grooved and either hinged or carrying a peg mechanism; they were spanned by muscular strength, for which the feet were probably often used. It seems likely from both the limited distribution of the crossbows and their type that they were copied from prototypes introduced by the Portuguese. This hypothesis is strengthened by a bronze figurine from Benin in the British Museum which shows a Portuguese armed with a crossbow and carrying at his waist what seems to be an iron ratchet (known in England as a rack, crick or gaffle) for spanning the bow.[31]

West African arrows consisted typically of a shaft into which an iron head was embedded by a long point or tang and then bound by thread. The pattern of the head varied considerably, in much the same way as that of the spears, and most recorded examples were barbed. Occasionally the point was obtained simply by burning and shaping the end of the shaft. The overall length of the arrows again showed many variations, the largest examined by the writer being one foot and so only a third of the length of the cloth-yard shaft of the longbow arrow. The shafts were of palm wood, savannah grass, or reed, and were sometimes fletched by feathers or stiff paper. They were generally packed into a locally made tubular quiver of leather or bamboo holding up to about fifty; for smaller arrows the young hollowed-out branches of the cabbage tree (*Anthocleista vogelii*) were sometimes used. The crossbow bolt (or 'quarrel') apparently differed little if at all from those used with the self bows; one example examined was rather lighter than most arrows, a reversal of the usual practice in Europe.

As with the heads of spears and javelins, arrows and bolts were usually dipped before battle into a poisonous brew of which the essential ingredient was *Strophantus hispidus*. This, if it entered the blood-stream, could cause death in ten to thirty minutes. Various antidotes, usually with alkaloid or glucose bases, were carried by troops.[32] Informants professed to believe that the poison retained its efficacy for many years and still handle bundles of old arrows with great care.

Secondary accounts of warfare in the Sokoto Caliphate include reports of the use of incendiary arrows. These would have been very effective against the ubiquitous thatch of African houses. Except for an account in a Hausa poem of the capture of Kanoma during the Fulani jihad in 1805, no first-hand description has been traced of any action when this weapon was used. Clapperton, however, was told of an aerial incendiary attack by Fulani against a group of Yoruba villages near Old Oyo when pigeons to which 'combustibles' had been tied were used to fire the thatch.[33] It seems possible that this somewhat tall story was a garbled or misunderstood account of an attack by incendiary arrows.

Probably the archer looked to his ruler or his chief to provide at least a part of his supply of arrows. It is known that at Benin and Yawuri between campaigns arrows were stored centrally in the royal arsenal,[34] a practice doubtless followed elsewhere.

The bow, like other weapons, has its symbolic counterpart. Among the Yoruba this takes the form of models, some only 10–12 inches in height, wrought in metal and often decorated in incised patterns.[35] These have been described by informants as insignia of office or rank or as associated with the rituals of specific cults and societies.

Simpler and doubtless still more ancient missiles used in West Africa for war as well as hunting were the sling and catapults by which pebbles or similar small objects were hurled against a target. Naturally enough, examples known to have been used in war have not survived,[36] but there are references to their use in tradition and prototypes may still be seen for hunting small animals in the bush.

Clubs and miscellaneous weapons

Probably the oldest, and certainly the simplest, weapon is the club. By the European Middle Age this had evolved into various forms, such as the axe, the mace, the hammer, and the halberd, which (with the exception of the last, a combined spear and battle-axe) were all hafted weapons with no point or cutting edge. In West Africa, in much the same way, such percussion weapons both retained some importance as secondary armament even into the age of firearms and also developed different and more sophisticated forms.

The basic and earliest weapon in the forests of West Africa, and in much of the woodland savannah too, can be assumed to have been a thick bough selected from a tree, having its branches trimmed and one end cut from the intersection of branches with the main stem to form the head. This could be used either as a cudgel or as a throwing stick. Bosman mentions the presence of such clubs in the Whydah army of the seventeenth century, every man having

five or six, all five or six inches thick; he adds that the
Whydah were very dangerous in throwing them.[37] Clubs
of this simplest kind are still found in compounds in the
West African countryside, where they are kept in readiness
for use against marauders. A Yoruba aphorism makes the
point that 'It is ridiculous to complain of having no weapon
when you are in a fight in the forest.'[38] It is sometimes
said that the heads of these weapons were smeared with
poison or with a magical substance.

From the basic wooden club have derived many other
forms, some, for example, studded with nails, others of iron
with iron coils wound round the shaft. Such weapons have
on occasions been transposed, like the mace in England,
into symbols of authority. Among the Yoruba the double-
headed axe of Shango, god of war, is primarily a ritual
instrument, though informants have claimed that warriors
were accustomed to sleep with such axes by their side so
that in case of surprise this should be the first weapon to
come to hand.

Among miscellaneous weapons, iron throwing knives –
especially those described by Barth as 'handbills'[39] – and
fighting bracelets were important. The latter were studded
with spikes for use against an adversary in hand-to-hand
fighting. Examples of both these are so far only known from
the savannah. Other devices used in war were pits covered
with leaves and lined at the bottom with spikes – an effective trap for cavalry – and calabashes filled with bees and
thrown as bombs.[40] Finally, cavalry made use of whips in
the pursuit of a fleeing enemy and for rounding up captives.
Idris Aloma's horsemen carried these in addition to their
swords and lances, but the Dagomba, according to one
source, sometimes disdained any weapon other than the
cowtail switches which they whirled around their heads.[41]

Armour and accoutrement

In the heavy cavalry of the Central Sudan the warriors, and usually their mounts too, wore armour. This was of two types: protective quilting, made of homespun cotton and stuffed with either kapok or wads of paper, and mail. The best account of West African armour as worn in action comes from Denham's description of the cavalry of Bornu. There the coats of mail, or hauberks, covered the horseman from his throat to his knees, dividing behind so as to protect the flanks of his charger. Alternative to the hauberk was a sort of quilted jupon which gave considerable protection against a sword slash and some protection against arrows; this was a garment sometimes also worn by Sokoto bowmen. Occasionally metal headgear with chin pieces was worn; more often hats were of straw or cloth, reinforced with leather and covered with a turban. Armour, either quilted or metal, was too hot, heavy and impeding for normal use and was donned only for action; Denham gives an interesting vignette of the chiefs of Bornu being fitted into their armour before the battle of Musfeia, their attendants working through the night with hammers so that the advance could begin before dawn. Light cavalry was less burdened, wearing cotton gowns over which mail coats were only occasionally placed.[42] Whereas the locally-made quilted armour offered adequate resistance to an arrow shot, and little more, mail gave greater protection against all pointed weapons. Yet despite his hauberk, Barca Gana, the early nineteenth-century commander of the Bornu army, was wounded in battle by Shuwa Arabs hurling light spears.[43]

Horse armour consisted of quilted covers for the chest and shoulders of the chargers and sometimes also metal face plates. Saddles were usually of the oriental type, with a high pommel and cantle to give protection and stability to the rider; they were usually placed over a colourful saddle-cloth. A simple bit of the bar or snaffle type, often in decorative brass, was used. Stirrups, again of brass, were shovel-shaped, usually decorated, and with their edges honed so that they could be used as an additional weapon and a spur; this type also provided a rest for the rider's lance or gun and could even act as a battering ram against earthen fortifications.[44] The less wealthy, however, made do with cumbersome wooden stirrups. As Law has suggested, it seems likely that horse accoutrement was introduced into West Africa during early Islamic times; significantly, the Hausa words for saddle, bit and stirrup all derive from Arabic. Similarly, many items in Yoruba equine vocabulary derive from Hausa.[45]

Armour was apparently little used in the Western Sudan, though the Mossi cavalry for protective purposes assumed as much clothing as possible and provided leather and copper shields for the vulnerable parts of their mounts. In the forest and coastal states, where horses were rare, it was also little known. It does not seem to have been part of the equipment of the Oyo cavalry, for example, though saddles and stirrups of the Sudanese (and ultimately Arab and Berber) patterns were in use in nineteenth century Yorubaland. Some chain mail was, however, possessed by kings and greater chiefs; two hauberks are preserved in the palace at Owo in south-eastern Yorubaland and tradition records the possession of metal shirts by an Oba of Benin in the late fifteenth or early sixteenth century and by an early chief at Oyo.[46] Dapper mentions a curious collar made of interlaced branches worn by the warriors of the Gold Coast to ward off blows,[47] but no examples of or other references to this primitive armour have been found. The leather war aprons of Yoruba chiefs and the padded war jackets of their infantry must also have given some protection against the weapons of their enemies.[48]

No reliable tradition of the manufacture or systematic repair of mail seems to be preserved in West Africa, and it can be assumed that the whole supply came from outside the region. The import of armour into the Central Sudan had begun by the sixteenth century. Leo Africanus describes the presentation by a man from Damietta of a hauberk to the ruler of 'Gaoga' – a kingdom somewhere to the east of Bornu – in or about 1512, and Ibn Fartua mentions the use of mail and also quilted armour in sixteenth-century Bornu, while according to tradition armour, both quilted and mail, was introduced into Kano as early as the reign of Sarki Kanajeji dan Yaji in the early fifteenth century.[49] The main source of imported mail was Egypt, much being probably second-hand Mamluk equipment, while until recent times there was an industry supplying the Sudan at Omdurman; Morocco could well have been another source, but no Maghribi armour has yet been recorded in West Africa.[50] Mail was also occasionally brought to the West African coast by European traders; an English list of merchandise suitable for trade in the Benin river at the end of the sixteenth century includes 'Shertts of mayle not many' among the items.[51]

Shields were carried by both cavalry and infantry. The usual materials for these, all locally-made, were hide, wood, and basket-work, while Bosman describes the shields of the Gold Coast as being sometimes covered with thin copper plate. They varied in size from small circular targes to the great shields of the Kanembu, each shaped (as Denham wrote) like a Gothic window and large enough to cover a soldier snugly as he lay on the ground in camp. These Kanembu shields were made of fogo wood, bound with hide thongs, and despite their size weighed only a few pounds. Rather oddly, the *Kano Chronicle* claims that shields were introduced there in the early eighteenth century from Nupe at the same time as muskets. But it seems reasonable to assume that shields of some kind to ward off arrows and sword thrusts have been in use everywhere almost as long as these basic weapons.[52]

Apart from their weapons, accoutrement and baggage, armies also carried with them less utilitarian impedimenta which were nevertheless of great importance for their morale. To the Yoruba, for example, their war standard, usually a staff covered with charms and amulets, was a symbol of their strength and also had a mysterious power of its own; sacrifices were offered to it before a campaign set out. The warriors of the Fulani jihad eschewed in theory such pagan devices, yet to each of the leaders of the expeditions which founded the several emirates, Usman dan Fodio presented a flag, which then became a symbol of authority as well as a rallying point.[53] Armies, both Muslim and non-Muslim, were accompanied by their own musicians, and soldiers advanced into action encouraged by the beating of drums and blowing of trumpets. These instruments were also used for passing orders, relaying intelligence, and confusing the enemy.[54]

Firearms; artillery; rockets

From the preceding account it will be observed that conventional weapons (as opposed to firearms) continued to play an effective role in West African warfare until as late as the middle of the last century. Yet the introduction of firearms into the region had taken place, though on a very small scale, some 400 years before, and by the end of the seventeenth century they had been widely adopted on the Gold and Slave Coasts, were beginning to penetrate the forest states, and had reached Bornu, Hausaland and elsewhere in the Sudan.[55] They came from two directions: the coast, where they had been brought by traders from Western

Europe, and across the Sahara, brought by traders from North Africa and the Near East.

The first firearms to be seen in West Africa were probably those carried by the Portuguese expeditionary forces who arrived on the coast in the last two decades of the fifteenth century. On political and religious grounds, and perhaps also because of their own dependence on Flemish and German gunsmiths,[56] the Portuguese Government forebade its nationals to sell firearms to the West Africans. This ban seems at first to have been successfully maintained in the case at least of Benin, and few guns reached that kingdom until late in the sixteenth century.[57] On the Gold Coast, however, the Portuguese found it necessary to arm the Africans living near their forts in order to protect them against hostile local peoples. The Elminas, for example, had bought guns from them and were adept at their use well before the end of the sixteenth century. In the seventeenth century the English, Dutch, Brandenburgers and Danes all began the sale of muskets.[58] Soon the Dutch became the major exporters of firearms to the coast, and though it was the Danes who gave their name to the flintlock musket which after 1750 was the principal weapon exported there, most of the 'Dane guns' in fact came from Holland. Although the trade in firearms never overtook that in cloth, guns and gunpowder were an important item in West European exports; they represented, for example about one-fifth of the value of cargoes shipped from England to Africa in the eighteenth century, and their purchase by the Africans was largely financed by the export of slaves.[59]

It was not until late in the seventeenth century that firearms began to reach the inland forest states from the south, at first only in small quantities. By about 1700, however, the Ashanti were well supplied, and it was their comparative strength in firearms which largely enabled them both to dominate their neighbours to the north, especially the

Dagomba, and to initiate their drive to the coast with its trading opportunities – especially of obtaining more guns and powder. It was noted as unusual that the Hausa company who aided them in their siege of Anomaba in 1807 (page 84 above) fought with bows and arrows as well as with muskets. Yet the distribution of firearms remained uneven. Apart from the Ekiti, who were in contact with Benin, the Yoruba seem not to have encountered them until 1726 when the Oyo cavalry overcame a Dahomean force armed with muskets. Though their horses were frightened by the noise of the enemy guns, they seem to have received no strong impression of the strength of this new weapon, and the Yoruba states made little or no use of guns until a hundred years later.[60] Again, guns were introduced into Dagomba (by the Ashanti) only in the late eighteenth or early nineteenth century, and were not adopted by armies in the Niger bend until the end of the nineteenth century.[61]

Firearms were not bought for military purposes only. Muskets were used in hunting and for the protection of crops, and for firing the *feux de joie* and other salutes which soon became popular; the small cannon which were imported along with muskets were probably mainly fired as salutes to great persons or on festivals or, as Bullfinch Lambe noticed in Dahomey, to mark the recurrent market days.[62] Nor were these weapons notably effective in battle. The flintlocks, though they had a range up to 200 yards – three times that of bows and arrows – were often wildly inaccurate at a mere fifty yards or so, required an interval of several minutes between volleys for reloading, and were probably more dangerous to the users than to the enemy since the barrels were prone to burst.[63] Nevertheless, by the mid-seventeenth century a specialized form of infantry, the musketeers, had arisen in those states where firearms were available in adequate quantities. Bosman reports that on the Gold Coast, where the Dutch were forced (as he claims)

by competition from the English, Danes and Branden-
burgers to sell 'incredible quantities' of firearms, the chief
weapons were now muskets, 'in the management of which
they [the local warriors] are wonderful dextrous ... they
handle their Arms so cleverly, discharging them several
ways, one sitting, the second creeping, or lying etc. that 'tis
really to be admired they never hurt one another'.[64] Other
writers, for example Captain Jones in describing the Egba,[65]
showed less enthusiasm for the West African skill in arms.
But the introduction of rifles and cartridges in the second
half of the nineteenth century led to improved marksmanship and rate of fire, important developments in the art of
war which were interrupted by the onset of the colonial era.

According to the *Kano Chronicle* guns had been introduced into Kano by a refugee from Bornu early in the
fifteenth century,[66] but the earliest clear mention of their
presence in the Central Sudan is in the account by Leo
Africanus, already cited on p. 105 above, of a presentation
made to the king of 'Gaoga' in or about 1512 by an
Egyptian visitor. The gifts included, apart from the nauberk,
a fine horse, miscellaneous weapons, and a handgun. The
first reference to the use of guns in this area, however, comes
from the reign in the late sixteenth century of Mai Idris
Aloma in Bornu who, so Ibn Fartua recounts, acquired
'Turkish musketeers and numerous household slaves skilled
in firing muskets.' They were probably obtained, Fisher considers, in normal trade between Bornu and Tripoli rather
than as a result of the Mai's pilgrimage through Ottoman
dominions to Mecca.[67] These firearms, which should perhaps be translated 'arquebuses' rather than 'muskets', were
used decisively in Idris's attack on the So town of
Amsaka,[68] yet Fisher and Rowland conclude that the Mai's
successes 'owed more to other devices than to guns'; Lavers
considers, too, that after their initial surprise value, it was
the 'mode of use' of the guns which counted, that is, the

organization of a disciplined corps of handgunners, rather than firepower. There are only two subsequent references to any supply of firearms to Bornu before the nineteenth century, these being accounts of presentations to the Mai's ambassadors by the Pasha of Tripoli in the early seventeenth century.[69] In Hausaland, according to the *Kano Chronicle*, cited by Hiskett, 'muskets began to be imported into Kano from Nupe in the reign of Muhaman Kumbari (1731–43), while Babba Zaki (1768–76) was the first to establish a bodyguard of musketeers'. Though muskets imported from the coast may possibly have been available in the Nupe markets as early as this, the report must be treated with caution. Firearms played little part in the Fulani jihad and were used sparingly, and as something of a novelty, in the Bornu wars of the early nineteenth century. As late as the midnineteenth century few were to be seen in Hausaland. Barth counted 'not more than four or five muskets' in a review of the Katsina cavalry in 1851, and he noted at Kano that very few firearms were to be bought there, 'although common muskets have begun to be imported by way of Nyffi [Nupe] at extraordinarily cheap prices by the Americans. Pistols and blunderbusses are privately sold by the merchants to princes or great men.'[70]

Various theories have been advanced for the apparent failure of the states of the Central and West Sudan to appreciate the potential value of firearms: for example, that their introduction was incompatible with the pattern of warfare, based on cavalry, practised in the region, or that this 'democratic' weapon threatened the ascendancy of the nobles (or, in particular, of the Fulani emirs of Hausaland). Smaldone considers that 'the only reason why firearms were not used extensively in the Central Sudan in the nineteenth century is that trading conditions did not permit the states of this area to obtain access to a source of guns.'[71] Yet an equally compelling reason – which applies to the period before the nineteenth century – may well have been that the military leaders had made a realistic assessment of the poor-quality muskets which were all that were available to them and found them not worth the prices which were being demanded.

Though Barth himself probably did not appreciate this, his remarks cited above show that the military situation in the Sudan was at the point of change at about the time of his travels in West Africa. During the second half of the nineteenth century firearms were being brought into the region in increasing quantities, by smugglers across the Sahara and, more important, by European traders now by-passing the coastal middlemen. Nupe, as Barth foreshadows, became, with its access to the Niger, a major market for guns and ammunition. Moreover, as the century moved into its last quarter, the quality of these weapons began to improve considerably in line with the advance made in European and American arms. This both stimulated the rise of standing armies in the way discussed in chapter IV above and had important consequences for the strategy and tactics of warfare which will be described in chapter VII below.

Until the introduction of breech-loading rifles towards the end of the pre-colonial period, the hand-guns which were imported into West Africa had all been muskets, that is, smooth-bore muzzle-loaders. These were successively of three main types: the matchlock (which may well be the type represented on the Benin bronzes), the wheel-lock, of which great numbers were never produced, and from about 1635 the flintlock, the last becoming the prototype of the numerous 'trade guns' or 'Dane guns' of West Africa. In the 1860s examples of the improved firearms already introduced into European and American armies began to reach West Africa; the Egba, for example, were said to have obtained a few breech-loaders in the latter stages of the

Ijaye war.[71A] Only in the mid-1880s, however, did the Dane-gun begin to be displaced. At this time the widespread adoption by the powers of repeaters made available for export to Africa a huge surplus of single-shot breech-loading rifles, obsolete in Europe but vastly more efficient than the Dane guns. Of the force which resisted the British advance on Ijebu Ode in 1892 a large number, perhaps a majority, were armed with Snider rifles of this type, and in the same year Samori's army fighting the French had some 3,000 to 4,500 rifles. The army of Rabeh, the adventurer from the Eastern Sudan who invaded Bornu in 1893, was said two years later to have 3,180 rifles and cap-guns; the latter (which were unrifled) predominated and were mostly double-barrelled.[72] Yet even today the flintlock musket (still called a 'Dane gun') remains in use for hunting.

A variant of the flintlock musket of which large numbers (possibly of no great age) are today offered for sale by the 'antique' dealers of Nigeria is the 'pistol' or, more accurately, short-barrelled musket or carbine. Except for those which Barth noted as being sold at Kano, pistols seem not to be mentioned in contemporary accounts of the armies and armament of pre-colonial West Africa, but their representation on carvings of the mounted warriors of Yorubaland suggests that their use may have been fairly widespread among cavalry in the later nineteenth century. Pistols were also in vogue as presents from European travellers to the more important of the West African rulers on whose help and hospitality they were dependent.[72A]

The European arms traders seem to have drawn a line (for one reason or another) at the supply of machine guns to the West Africans; probably prices were prohibitive and the supply too limited. The writer knows of only two possible instances of the use of this weapon in inter-African wars in the pre-colonial period: the tradition that a Gatling preserved at Buguma in the Niger Delta was used in the

Kalabari civil war of 1879–83 and the allegation that the Ijesha were using a Gatling against Ibadan at Kiriji during the early 1880s.[73] But there may have been attempts long before then to mount guns in series on war canoes. Landolphe describes with much detail a curious arrangement which he claims to have seen on the war canoes of the king of Warri on the Benin river early in the nineteenth century; this 'jeu d'orgues' consisted of seven blunderbusses mounted in sequence on a swivel which could then be fired simultaneously. He adds that this weapon 'is rarely used'.[74]

Turning to the accoutrement of the West African musketeer, this included a flask containing the powder used both for priming and for firing, and a leather bag with a supply of bullets. The Dahomeans in the mid-nineteenth century, however, and possibly other troops, were using cartouches – made-up charges in paper packets – which they carried in belts slung over the shoulder or tied round the waist.[75] Presumably a stick for ramming the powder was also standard equipment.

The use of artillery did not develop in pre-colonial West Africa to the extent of that of hand-guns, and its handling in war was less well understood than that of the muskets. Those larger calibre guns which were imported were mostly brass cannon of the four- or six-pounder types. Bosman writes that there were a few cannon on the Gold Coast in his time and that they were very popular for firing salutes.[76] Such guns were occasionally placed on locally-made wooden carriages, these being perhaps the first wheeled vehicles made by West Africans.[77] They were also frequently mounted on pivots in the prows of canoes, and they featured to a greater extent in river and lagoon than in land warfare;[78] indeed, the canoe, with its ability to absorb the recoil of the gun in the water, must have seemed the ideal medium. A few cannon penetrated to the West Sudan, probably mostly from the south via the Nupe market, though the two four-pounders in the Bornu army for which Denham's carpenter constructed carriages, and which were used effectively at Angala against the Bagirmi, had been sent as presents from the Sultan of Fezzan.[79] To the west El Hadj Umar, founder of the short-lived Tukulor empire of the mid-nineteenth century, made good use of captured French artillery against rebels in his own territory. His son and successor, Ahmadu, made strenuous but unsuccessful efforts to obtain more guns from the French. Rabeh's army in Bornu in 1895 had forty-four pieces of artillery, described as 'small, inferior, unrifled, mounted on stands, but (with) no wheels'.[80] In the forest the Ijebu are said to have had a few cannon early in the nineteenth century, and at mid-century seven 'field pieces' were presented by the British Government to the Egba of Abeokuta at the instance of the Christian missionaries. Despite the lessons in gunnery which the Abeokutans received from Commander Forbes R.N. in 1851, their guns soon became no more than 'a roost for cocks and hens' since 'they eat too much powder'.[81] Cannon, including 'long carronades', were brought in quantities to Dahomey, where they then suffered the usual neglect,[82] and also to Lagos where the British attack of 1851 was strongly resisted by over forty guns of miscellaneous types – one with a barrel over ten feet in length – mounted in batteries defended by an earthen wall, a trench and stockades along the whole waterfront.[83]

The introduction of firearms into West Africa, prolonged as it was over several centuries, was far from revolutionizing warfare. It certainly seems unlikely, before the advent of the rifle, to have increased enemy casualties. Barth describes a whole Bornu army opening fire on groups of Musgo pagans with almost no success. 'But African muskets are not exactly like Minié rifles', he adds, 'and a musketeer very often misses his aim at a distance of thirty or forty yards'.[84] This comparative ineffectiveness of the firearms used in West African wars was not wholly due to the primitive nature of the weapons, but also to a lack of skill and method in their firing, to lack of maintenance, and to poor quality powder and ammunition. Most West African musketeers crammed powder halfway up the barrels of their guns, packed them with banana fibre or other forms of wadding, and then – expecting the worst, from sparks and recoil or exploding barrels – fired at arm's length or from the hip. Only the Ashanti and, according to Burton, the Amazons of Daho-mey, fired from the shoulder and sighted their weapons. The former, Dupuis writes, were 'trained to firing with celerity as we ourselves use the musquet'. Moreover, apart from one indication in the passage from Bosman cited above (p. 109) and the well-trained *sofa* of Samori's armies at the very end of the period, it seems that neither rifles nor muskets were fired from the prone position despite its advantages of cover and steadiness. As late as 1892 a large Ijebu army equipped with numerous Snider rifles and abundant cartridges was unable to make effective use of these against the small British-led force advancing against them through the bush.[85]

An interesting comparison is afforded by the situation in East Africa. Here firearms spread even more slowly than in the west. Such martial peoples as the Masai and (prob-ably) the Ngoni never adopted them. Yet this did not detract from their military superiority over their less warlike and disciplined neighbours. Similarly, the troops of war-lords like Mirambo were able with their spears alone to play havoc with the gun-protected caravans of the Arab traders.[85A]

The cheap muskets imported in such quantities into West Africa had a useful life of only some two to three years. As already indicated at the beginning of this chapter (p. 91), some repair of firearms was carried out by local artificers, the effectiveness of which probably varied greatly. Though stocks were easily made and fitted, it is doubtful if much could be done to replace or improve worn-out firing mechanisms and barrels. With one, very late, exception, no significant manufacture of hand-guns seems to have taken place, despite some claims to this, as also to the casting of cannon.[86] The Ashanti blacksmiths did succeed in producing brass blunderbusses, but these were useless for war; they were known as *humu* guns from their absurdly large barrels.[87] Landolphe reports that the Oyo 'ambassadors' whom he met on the Benin river early in the nineteenth century told him that, in their country, foundries existed for the casting of cannon and other arms, but this claim – which may have resulted from a misunderstanding or been the easy boast of men far from home – must be discounted for want of corroboration, as must also the statement by king Behanzin of Dahomey to a French mission in 1891 that an 'armament factory' in his country was producing firearms and powder 'according to methods learned from Europeans'.[88] The exception is again provided by Samori, the creator of the Dyula empire. In the 1880s he concentrated his ironworkers in the village of Tete where they succeeded in manufacturing flintlocks at a cost lower than the price paid for those bought from Freetown. Tete was evacuated in 1892 but the armament workers, said to number 300–400, were reassembled at Dabakola. By then they had succeeded, under the direction of an artificer who had spent several months in a French arsenal, in making effective copies of Kropatschek repeating rifles, eventually producing two of these guns daily.[89]

Bullets were manufactured locally (as they still are), though when rifles began to be imported cartridges were introduced with them. Iron bullets were believed to be more effective than those of lead. Burton describes the bullets which he saw being made in Dahomey as 'badly fitting bits of cut bar, sub-circular and all facets; they must fly wide and they cannot hit hard', and Barth mentions 'pewter' bullets harmlessly pattering against the wickerwork shields of the Musgo.[90]

The gunpowder used in West Africa was for the most part imported, Danish and Dutch powder being preferred to English, and English to American and French, though in fact poor powder, with a low saltpetre content, was safer when used in the weak and often unproved barrels of the trade guns. Local manufacture was hindered by a shortage of sulphur, but in the Gold Coast states at least a part of the supply was home-made. No differentiation seems to have been made between the powder used for priming the pan (which should normally be fine) and that for the propellant (for which coarse 'corned' powder, in grain form, was required). Control of the import of powder was as vital a political and military objective as control over arms imports; a quarrel over a consignment from the coast was the occasion of the Sixteen Years War in Yorubaland. In Upper Volta in the late nineteenth century powder was relatively more costly than guns.[91] Flints were imported with the flint-locks, but since a flint could be used only for some twenty to fifty shots, spares were needed. Until about 1800 most flints probably came from France, but after that time Brandon in East Anglia obtained a near monopoly of the West African trade.

The use of rockets as weapons was demonstrated to the West Africans on several occasions during the nineteenth century. Clapperton, for example, set off three rockets, fixed to spears, for the benefit of the Sheikh of Bornu, with whom he had been discussing methods of attacking a walled town; they were then used with effect during the Sheikh's action against the Munga rebels.[92] Rockets were also used by the British against the Ashanti at Dodawa in 1826, by Consul Foote of Lagos against Porto Novo in 1861, by Governor Glover of Lagos against the Egba at Ikorodu in 1865, and by the British expedition against the Ijebu at Imagbon in

1892. These actions showed the effectiveness of rockets in setting fire to dry bush and to thatched roofs, and though they produced few casualties their noise and appearance caused greater terror than did gunfire. A few West African armies possessed and used rockets independently of the Europeans: the Nupe were said to have had a number in 1871 while the Ekiti at about the same time obtained what proved to be signal rockets; an Ijesha proverb complains of the ineffectiveness of the 'shooting stars' used by one of their war leaders.[93]

Training

European observers of the West African scene from the seventeenth century onwards, conditioned by their experience of the comparatively well-drilled and disciplined armies of their own countries, usually emphasize in their accounts the apparent lack of order and professional expertise among the officers and troops of the countries they visited. This attitude culminates in the sardonic and lofty descriptions by Richard Burton of the Dahomean army, echoed in Captain Jones's disparaging account of the Egba operations at Ijaye. Where praise was given, as for example by Dapper for the marksmanship of the Elmina, there is a sense of surprise and patronage. Allowance must be made for ethnic prejudice, but even so there remains substance in the charge that the West Africans failed to make the best use either of their manpower or of their weapons. This was indeed a major factor in their collapse before the small colonial expeditions sent against them in the late nineteenth century,[94] and it also accounts for the indecisive nature of much of the action in their own inter-African wars.

This ineptitude may be ascribed as much to innate conservatism as to lack of military training, for even before the rise of standing armies in parts of West Africa towards the

end of the period, training of a sort existed. The general
liability for service in war which existed among most people
did not find the young warrior unprepared. His boyhood
had been enlivened by games, encouraged and overseen by
his elders, which simulated war, and his manhood was
entered upon through rites which were intended to innure
him to the bloodiness and hardship of war.[95] Fighting
peoples took care too that their sons were early initiated
into real combat. Snelgrave noted at Abomey that the
warriors in the royal army were followed by boys who
carried their shields, and he was told 'That the King
allowed every common Soldier a Boy at the publick charge,
in order to be trained up in Hardship; and that the greatest
part of the present Army consisted of Soldiers, bred in this
manner.' A somewhat similar institution existed in Ibadan,
where cadets were taken by their elders to the battlefield.[96]
Such exposure to the danger and reality of war provided
the most effective training.

In addition to this training of the young, reviews and
manoeuvres simulating war were undertaken by the armies.
Labat, for example, describes the somewhat disorderly exercises performed by the cavalry of Futa Toro in the eighteenth century. Accounts in the next century show the
infantry of the King of Dahomey practising attacks on an
imaginary enemy, while others describe the armies of Sokoto
and Bornu parading for review; Denham was astonished by
the 'tact and management' of Bornu horsemen on one such
occasion.[97] That such exercises were not confined to the land
is suggested by the manoeuvres which were performed for
the edification of the British Consul and a French official
guest in 1861 by the Lagos fleet of war canoes.[98]

Arms drill and weapon practice can be assumed to have
had a place in training, though few references to this have
been found. Lander's report (mentioned on p. 98 above)
of the marksmanship of the Oyo archers implies regular

practice in the use of this somewhat specialized weapon. Firearms required even greater skill and practice. The Kambonse, or musketeers, of Dagomba were said to undergo protracted training in the use of their weapons, and their officers were attached for instruction to the army headquarters at Zugu, a Dagomban Aldershot near Yendi.[99] But this was exceptional and it appears that in the nineteenth century powder was too scarce and expensive and real war was in any case too frequent for West Africans to pay the sort of attention to practice firing and other training which these received in other armies. Moreover, serious training in the use of firearms could most suitably be undertaken by full-time soldiers (as instructors of other, preferably also full-time, soldiers), and it was not until the last decades of the pre-colonial period that standing armies of regular troops developed. Only with this, and with a consequential improvement in discipline and the handling of weapons, could the potential strength of the West African states be mobilized and, before there had been time for this to come to pass, the armies of the colonial powers had overcome all organized resistance to their advance.

Horses and other animals used in war

The horse and the canoe are not usually thought of as weapons, but their use in war makes that description appropriate. Both were swift and efficient means for transporting warriors into action; the first was also a cavalry charger, a light cavalry mount, and the bearer of commanders and their orderlies, while the second was likewise in itself a unit of battle. Both gave mobility to the armies of West Africa, the horse on land, like the tanks of the twentieth century, the canoe on the placid waters of rivers and lagoons, a humble forerunner and relative of the ironclad navies of the seas. Of these two adjuncts to war, the horse will be treated first.

Humphrey Fisher considers that horses had become established in the Central Sudan well before the period when written records for the region begin – about 1000 A.D. – and thus before the coming of the nomadic invaders who are often credited with the founding of such states as Kanem and the Hausa kingdoms and also before the arrival of Islam.[100] From the first horses, very early arrivals presumably from across the Sahara, have descended, Fisher suggests, the small horses which have been persistently remarked as present in many parts of West Africa from the eleventh century (in Ghana) onwards and which Fisher terms the 'Southern Sudanic breed'.[101] From the Central Sudan this breed spread into the Western Sudan and southwards into the forests and to the coast. In addition to these indigenized horses (or ponies, as Europeans called them), there has been from the fourteenth century onwards an import of larger horses from North Africa. This, first mentioned by al-Umari in his account of Mali, seems to have brought about the development of cavalry, 'something of a revolution in West African warfare' in Law's words, and to have been linked with the introduction about the same time of the saddle and stirrup and of armour.[102] At Timbuktu Leo Africanus noted that: 'Here are verie few horses, and the merchants and couriers keep certaine little nags which they use to travell upon: but the best horses are brought out of Barbarie.' Such horses were bought in Europe for ten ducats and sold in Songhay for forty or fifty. Similarly, 'the King of Bornu sent for the merchants of Barbary, and willed them to bring him great store of horses: for in this country they used to exchange horses for slaves, and to give fifteene, and sometimes twentie slaves for one horse'[103] – a rate which varied little during the next three centuries. On the coast south of the Senegal river the Wolof in the late fifteenth and early sixteenth centuries were buying horses from Portuguese traders as well as from the interior.

Fernandes adds that their king, who was 'lord of 8000 horsemen', bought his mounts 'less for war than for his own honour'. The use of cavalry by the Oyo seems to have developed in the latter half of the sixteenth century, though they may have used horses earlier for non-military purposes.[104]

In some parts of the Sudan horse breeding was possible, and as a result of this there grew up an east-west trade which possibly rivalled the Saharan trade. Particularly famous were the horses reared in the Mandara country east of Adamawa; these, often described as Bornu horses, have been named by Fisher the 'Bornu/Mandara breed'. Horses from this area were the mainstay of the armies of the Sokoto Caliphate, which had adopted the use of cavalry in imitation of their Habe adversaries, and probably also of the Oyo cavalry in its heyday before the end of the eighteenth century.[105]

References to the two main types of horses, the small local breed found over much of West Africa, and the larger ones, either imported or bred in the eastern parts of the Central Sudan, continued to be made by travellers in West Africa down to the end of the nineteenth century. At a horse race in Borgu, for example, Clapperton noted that: 'The horses run in pairs ... sometimes a large Bornu horse paired with a small native breed, the latter of which appeared to dispute the victory often with the larger horses of Bornou'.[106] But, as Law remarks, even the larger breed was small by European standards.

The maintenance of horses in West Africa was peculiarly difficult because of the prevalence over wide, and often shifting areas of the tsetse fly, as well as lesser-known carriers of disease. This, combined with the long gestation period, made breeding even more difficult. Yet in the savannah many horses seem to have acquired some immunity to the types of fly found there (some by contracting disease and recovering), and it was probably this which allowed the evolution of Fisher's Southern Sudanic horses, while the Mandara hills, home of fine horses, may have been altogether free of tsetse. In the forest margins to the south horses were very vulnerable to two types of tsetse, the *Glossina longipalpis* and the *Glossina palpalis*, which enables Ojo to claim that 'the tsetse fly preserved the integrity of the Yoruba culture against forces originating from the north as effectively as the mosquito did for many decades in the south'.[107] Moreover, it seems that the extension of farming in the forest during the nineteenth century, and especially the cultivation of the oil palm, may have considerably increased the number of flies in the area.[107A] Some protection was possible against the flies, various 'medicines' being prescribed and also the keeping of smoky fires near the stabling, but in general, as the accounts of travellers show, the expectation of life for the horse in most parts of West Africa was low. The average life of a horse in Samori's cavalry, for example, was estimated at no more than about six months, though hard-riding as well as disease contributed to this.[108]

Apart from disease, the lack of convenient fodder was a reason for the scarcity of horses outside the savannah. Grazing was hardly possible in the forest so that suitable grasses and grains had to be collected. In Gonja, for example, a chief owning a horse assigned one of his wives and a boy to look after it, and this was doubtless the pattern elsewhere. To the Yoruba, a man who had a horse but employed no groom typified someone who was living beyond his means.[109]

Mention has already been made of the Arab-type saddles used by the cavalry of the Sudan and which penetrated also to the Oyo. But not all horsemen of the region used saddles. The Jukun, Bedde, Musgo, and Angas of the Central Sudan were all expert bareback riders. The last three, it is said, riding naked, habitually slashed the backs of their mounts so as to glue themselves in their seats by the congealed blood.[110]

As well as horses, there was some use of camels by the armies of the West and Central Sudan. According to Ibn Fartua, these were introduced into the Bornu army in the sixteenth century by Idris Aloma who ordered his chiefs to procure them. They supplanted, or more probably supplemented, the droves of oxen, ponies, mules, and donkeys previously used for transport in Bornu, and enabled the Mai to send long-range expeditions into northern Kanem, where they also took a part in battle; Ibn Fartua praises the camel corps, 'consisting of Barbars and Koyam', who at Milmila hill 'did not dismount in spite of the enemy's furious onset'.[111] Camels were also well-adapted to scouting and skirmishing in the desert and semi-desert parts of the region as well as to carrying baggage, and in the Sokoto Caliphate they were used by the nomadic Fulani and the Tuareg mercenaries who acted as pathfinders to the army.[112] Horses usually exhibited great fear of camels, so that where conditions of terrain and climate were suitable for their use, camels could play an important role in battle against cavalry, despite their comparative lack of speed. Thus the desert Tuareg were sought-after allies in the wars of the Sudan.[113]

War canoes and transports

For the riparian and lagoon- and lake-side peoples of West Africa boats were of as much importance in war as horses for those in the savannah. The most familiar and widely distributed type of boat was the dug-out canoe,[114] found both along the coast and on inland waters. It was made by felling a selected tree and then trimming, excavating and 'carving' the trunk with iron tools, a process often completed by burning out the interior and then sometimes also soaking the canoe in water for some weeks. The dug-out varied greatly in size, from the one- or two-man shell to vessels up to 80 or even 100 feet in length and capable of carrying some 100 men with their provisions. The warlike Iso, living west of Badagry, fought with their spears from the smallest kinds of canoe, but usually the war canoe was of medium size – the largest, of course, lacked stability. Canoe building was a specialized craft, and though most canoe users made at least some of their own boats, there were recognized centres of the industry. In the Niger delta, for example, where the ubiquitous mangrove was unsuitable for boats, many canoes were supplied by the Ijo from the forests of the upper Benin river.[115] Songhay obtained some of its boats from Kebbi, as is known from a letter which the Sultan of Morocco sent to the Kanta of Kebbi in or about 1594 demanding that he should 'hand over (to my forces) the whole of the annual quota of boats which you used to give to Askia'; these boats were probably built for the Kebbi by the Sorko living along the Niger.[116]

In the savannah another variety of boat was not uncommon: this was the locally-made boat of planks (presumably carvel- not clinker-built). Ibn Fartua implies that it was boats of this type which Idris Aloma of Bornu devised for river crossings by his armies. Labat also mentions small plank-built boats as being in use on the Senegal River in the early eighteenth century; in the nineteenth century Denham saw plank-built boats nearly fifty feet long at Logone on the Shari river, and Barth saw them both on the Niger and on Lake Chad, those on the Niger being 'sewed or tied together in a very bungling manner'.[117] The Yedina warriors of Lake Chad had curious small craft like surf-boards, made with planks cut from the light fogo wood of the ambach tree (which was also used for shields), but for their piratical expeditions they used their famous papyrus canoes.[118] In the south of the region a few plank-built boats were probably bought from European ships,[119] while conversely Europeans on the Gold Coast obtained dug-outs from such noted builders as the men of 'Achombene' and Takoradi.[120]

VI. Static warfare and fortifications

Much of the warfare of pre-colonial West Africa was of a static character and consisted largely of the investment by the army of one people of the military and political centres either of another people or of a rebellious tributary. This view derives mainly from accounts by European observers in the nineteenth century, but earlier evidence suggests that sieges and the formation of permanent and semi-permanent camps as military bases in enemy territory were long-standing and prominent features of West African warfare. The evidence comes not only from the wars of the peoples of the forest, where physical conditions hindered pitched battle and manoeuvre, but also from those of the savannah, a land better adapted to the mobile warfare of cavalry and to the use of long-range weapons.

The widespread and numerous remains of fortifications, in particular the mud-built walls around towns, are a recurring and sometimes conspicuous feature in the West African landscape. The walls fulfilled purposes additional to those of providing a means of defence. They served notably as boundaries, demarcating a town from its farmland, regulating trade, and enabling tolls to be collected from travellers. Thus many West African town walls as observed today follow a course which seems to ignore tactical opportunities offered by the nature of the ground. Nevertheless, what can be learnt or deduced about the evolution of towns strongly bears out the common-sense assumption that in origin the walls were of military significance, providing defence against an enemy and shelter for the inhabitants. There is abundant evidence too that walls continued to be the salient features in defensive warfare. That in most cases they covered what seemed from the tactical point of view a dangerously large area was noticed by several European travellers, but the open farmland and gardens enclosed within them served to feed a town during a siege and also enabled the defender's army to be drawn up and rallied for battle behind them. That many, probably most, towns originated as military strongholds cannot be proved, but there is much evidence for such a contention both in the sites chosen and in the recorded history of some comparatively recently founded towns such as Sokoto, which was first inhabited in 1809 as a camp formed by Muhammad Bello, and Ibadan, which evolved from the camp of the shattered Oyo armies about 1830 and is strongly placed on a ridge of hills near the edge of the forest. Only one West African case has been noted of a wall which surrounded not a town but also its dependent farmland. This is the great earthwork known as the Eredo which, in a circuit of some eighty miles, enclosed the Yoruba kingdom of Ijebu Ode.[1]

The formation of a fortified camp, distinct from the parent town or towns, was usually the first step taken by a West African army when it advanced into the field. Where camps were formed on the march, they were naturally of a temporary character; the leaders were sheltered by tents or by walls of matting while the soldiers slept under such shelter as they could find – in the case of the Kanembu spearmen, under their great shields – or in the open. But on arrival at the point chosen by the commander as the base of operations, the practice was to throw up an earthen wall surrounded by a ditch (the excavation from which the wall had been built). Camp followers and auxiliaries added to the population and semi-permanent quarters for all were quickly erected according to the varying local patterns.

Soon a market developed within the perimeter and if the occupation was of long enough duration – as was usual – the soldiers and other inhabitants resorted to farming. Thus a West African military camp acquired the appearance of an ordinary village. Smaller permanent camps were also part of the defensive system practised by West African powers, especially with the object of consolidating territory gained from an enemy or to parry a threat to the centres of power at home. Examples of these are the 'stockades' which the people of Kano planted in the sixteenth century along the Bornu marches, either for defence or, as Ibn Fartua alleged, as part of their aggression against Bornu; the numerous 'war-towns' of the Mende and other peoples in Sierra Leone; the *ribats* which defended the frontiers of the Sokoto Caliphate, and the fortified outposts of Yoruba towns such as Ibadan.[2]

For the most part, these West African fortifications, whether for the protection of towns or of camps, consisted in roughly shaped earthen walls. They were usually continuous and elliptical in plan, the more important towns – in particular, those in the forest – being defended with two or even three circuits. Two main types of such walls may be distinguished, the first being massive constructions up to thirty-five or forty feet in height, and the second being breastworks some four or five feet in height. Free-standing walls of earth are also occasionally found, often adjacent to gatehouses (whose walls were naturally also free-standing) and sometimes as parapets; these were generally pierced with loopholes for archers.[3] At Kano, and probably other towns of the savannah, earthen ramparts were fronted by free-standing walls of which the hand-shaped bricks (*tubali*) were faced with a mud cement which was renewed annually after the rains. Elsewhere palisades were constructed of sharpened stakes and bamboo. These were characteristic of waterside towns but were also found inland, both in the forest and the savannah, while sometimes palisades were erected (as apparently at Ile Ife) along the top of a mud wall. Such methods of fortification were combined at the capital of the small kingdom of Cape Mount which was 'walled by balks, placed horizontally in two rows about three feet distant, the interspace being filled with clay'.[4] There are also a few examples of stone being used for defensive walls.[5] Those in the Yoruba hill country appear from their remains to have been of the breastwork type and of rough construction, but the stone walls of Surame, the capital of the Hausa kingdom of Kebbi, were much more substantial. E. J. Arnett, who examined their remains, wrote that: 'The striking feature of the walls and whole ruins is the extensive use of stone and *tsukuwa* or very hard red building mud, evidently brought from a distance. ... The walls show regular courses of masonry to a height of 20 feet and more in several places.'[6] Finally, a simpler, but effective, defence was a hedge of cactus such as surrounded, for example, the towns of the Kagoro, growing to some fifteen feet in height and of labyrinthine complexity.[7]

The gateways set in the defensive walls were often constructions of elaboration and subtlety, the entry contrived so as to include one or more right-angled turns;[8] they sometimes included galleries or platforms from which archers and, later, musketeers could direct enfilade fire upon an attacking force.[9] The gates themselves were usually of stout planking, sometimes covered with sheet iron as at Kano and Willighi, and occasionally wholly of iron, as at Bauchi and Birnin Naya; some of the gates at Kano were, however, found by the British expedition of 1903 to be merely of cow-hide.[10] Towers, designed for both defence and observation, were sometimes placed along the walls, like the 'turrets' which Jobson saw in Mende country, the watch-towers placed at the angles of the walls of Willighi in Bornu, and the thatched galleries raised on wooden stilts which the Egba built to defend their camps at Ado and Ijaye.[11] Within the towns, the principal houses and compounds, and especially that of the ruler himself, were also usually walled and provided further centres of defence.[12]

The building of these walls must have required a considerable community effort, though possibly rather less than might at first be supposed; Connah has estimated that the construction of the great inner wall at Benin would have required the continuous employment of 5,000 men for ten hours a day throughout one dry season.[13] The usual method was apparently to excavate earth by digging a ditch on the outside of the projected wall and then, after puddling the earth with water,[14] to pile up layer upon layer of mud. This was roughly shaped as the work proceeded from a broad base up to a narrow parapet, and formed, in the archaeologist's jargon, a 'dump rampart' – though the term 'wall' for such constructions has the sanction of very long usage.[15] Clapperton gives an interesting glimpse of the method by which the walls of Boussa, in Borgu, were kept in repair and the task of doing so organized:

> Bands of male and female slaves, accompanied by drums and flutes, and singing in chorus, were passing to and from the river with water, to mix the clay they were building with. Each great man has his part of the wall to build, like the Jews when they built the walls of Jerusalem, every one opposite his own house.[16]

Those parts of the walls which were not 'dump ramparts' but free-standing were constructed of hand-moulded bricks dried in the sun and cemented by the addition of more mud, as in the building of houses.[17]

Graham Connah has argued persuasively that 'In general the built wall [in West Africa] seems to be a savannah form, whereas the "dump rampart" is a high-forest one. The basic reason for the difference is probably rainfall', free-standing walls not surviving long in the heavy rains of the forest.[18] Much more research into and classification of West African defence works is needed before this generalization can be accepted. The evidence so far shows that in fact both types of construction were practised in both habitats, though it does also suggest that the free-standing wall was more prevalent in the savannah. Whether in the forest or the savannah, the material used in these walls and the frequent rebuilding and additions to which they have been subjected make it difficult to attempt even the most approximate dating, though sometimes (as at Ife) a *terminus a quo* can be obtained by dating material or objects uncovered at the base of a wall.

In numerous cases, as for example at Benin and the more important Yoruba towns, a double or triple circuit of walls was built, the innermost wall usually being the most massive.[19] The interval between the circuits was considerable, ranging from 100 yards to about a quarter of a mile. This distinguishes them from the fortifications of ancient and medieval towns in Europe and Asia and reflects a different tactical intention, the intervening spaces being used not so much to forbid the entry of an enemy as to extend the immediate jurisdiction of a town while still protecting the inner area and to give advanced shelter to an army being drawn up for battle to take place beyond the walls – as at Oshogbo in *c*. 1840.

Fronting the walls on the enemy side there was usually a ditch,[20] dug to a depth commensurate with that of the wall and either increasing the total height of the wall by continuing the slope without interruption or separated from it by a narrow berm. The ditches were converted by rains into moats, while their passage in the dry season, when most campaigns took place, was obstructed by stakes or by thorns planted deliberately along the bed. At points where gates were set into the walls, the ditch was crossed by bridges, usually of easily removeable planks. Beyond the outer walls it was customary to leave the immediate vicinity uncultivated, so that, in the forest regions especially, an enemy force had to approach a town by narrow paths twisting through thickly growing trees and undergrowth where ambushes could be conveniently laid.

There has been some debate as to the intentions which prompted the builders of the two main types of walls prevalent in West Africa, the massive wall of twenty feet or so in height and the breastwork. One suggestion is that the former was designed to obstruct the advance of cavalry and that such a scale was unnecessary where a cavalry attack was unlikely to be made,[21] while a second suggestion is that the lower wall was introduced later than the massive one – probably in the late eighteenth or early nineteenth century – as a support for musketeers.[22] These are both attractive theories, but they have to be reconciled with the existence of the two types of wall in both the savannah and the cavalry-inhibiting forest and with the lack of any positive evidence that the breastworks were of later date. Moreover, as has been pointed out above (p. 115), West African musketeers almost always fired from a standing position, even when fighting along their walls, as at Abeokuta in 1851. More important factors in determining the size of the walls to be built seem likely to have been the availability of manpower and the degree of urgency in the construction of defences. Paradoxically, the massive walls were probably in most cases the product of stable times while the lower walls were flung up hastily when enemies threatened – as in the case of the triple breastwork round Ijaye, the capital of Kurunmi's short-lived state in mid-nineteenth-century Yorubaland.

The construction of even a modest wall and ditch fortification around a town or village occasioned great labour, and most villages and many small towns in West Africa remained unprotected in this way. But where more important towns were unwalled, this was seen as exceptional. Leo Africanus, for example, thought it deserving of comment that Gao, the Songhay capital, had no walls. Clapperton mentions that Garbua on the Hausa-Bornu marches was ironically so-called – 'the strong town' – because in fact it was unwalled. Again, Bowdich observed that the Ashanti had no fortifications. The Dahomean capital, Abomey, though 'a very large town', had 'no wall nor breastwork to defend the besieged, nor are there springs of water in it'; it was, however, surrounded by a deep moat, crossed by four wooden bridges at each of which guard-houses were placed. Burton adds to the information given by Norris and Dalzel that the ditch was planted with thorny acacia and that the Kana gate was screened by a mud-built wall about 100 yards long.[23]

Professor Abdullahi Smith associates the rise of the Hausa kingdoms with the evolution of the *gari*, or township, into the fortified *birni*, or capital.[24] He laments, however, that 'Because of the fact that no archaeological investigation of *birni*-type sites in Hausaland has yet been undertaken we cannot indeed speak with precision about their antiquity.' This is the case too with the towns of Kanem-Bornu.[25] The deficiency is especially unfortunate since the savannah fortifications have enjoyed more favourable conditions for their preservation than those of the south where prolonged and violent rainstorms have eroded the earth and mud-brick constructions. Moreover, an excellent starting-point for such investigation would lie in the detailed allusions to savannah fortifications by Denham, Clapperton, Barth and other travellers. The account of the walls of Kano published by Moody in 1967 constitutes an exception. As Moody shows, the Kano walls underwent many changes throughout the duration of their usefulness, a feature in which they are undoubtedly typical. These changes are difficult to date because of the nature of the material, but Moody advances a plausible theory that the 'earlier walls' of Kano were wholly built as a dump rampart and without the use of moulded bricks.[26]

So far as Hausaland and its immediate neighbours are concerned, the jihad of Usman dan Fodio, beginning in 1804, makes a landmark in the history of fortifications, all the authorities stressing the encouragement that the leaders of the jihad gave to the building of walls around towns and permanent camps – the walls of Gwandu, for example, were under construction while Sheikh Usman lay ill within that place.[27] The fortifications of many Hausa towns, however, are of considerably greater antiquity than the age of the jihad; the first wall at Kano, for example, already enclosing the twin hills of Dalla and Goron Dutse, can be ascribed fairly confidently to the twelfth century, while an important extension was added in the fifteenth century under the famous ruler, Muhammad Rumfa.[28] Leaving aside the archaeological problems which they raise, these Hausa walls are still impressive, but in the mid-nineteenth century, when kept in repair, were much more so. Barth, surprised by their scale, estimated the Kano walls at between thirty and fifty feet in height, twenty and forty feet in breadth, and probably over fifteen miles in circumference, with thirteen gates; he gives similar accounts of the walls at Katsina and elsewhere.[29]

The antiquity of the system of establishing fortified camps as army bases during a campaign appears from Ibn Fartua's account of the wars of Idris Aloma of Bornu, in particular the foundation of the Sansana al Kebir, the 'Great Camp', in preparation for the conquest of Ngafata; similarly, in his war against the 'obdurate' Talata who had 'vaunted their shining white spears', the Mai set up a series of ribats in which he settled his Muslim warriors 'so as to allow ... little empty country' to the enemy.[30] The system was further developed during the jihad of the early nineteenth century, Muhammad Bello founding many ribats on the frontiers of the Sokoto Caliphate, especially against the raiding Gobirawa and to a lesser extent the Tuareg and the men of Kebbi. Murray Last suggests that in this Bello was consciously following methods adopted by the Muslims in their Persian, Syrian and North African conquests,[31] but as we have seen, there was already a strong tradition of fortified camps in advanced positions in West Africa itself.

In Bello's camp to the east of Sokoto, Clapperton, in a rather contradictory passage, notes that the troops 'take up their ground according to the situation of the provinces, east, west, north, or south; but all are otherwise huddled together, without the least regularity', continuing that 'The man next in rank to the governor of each province has his tent placed nearest to him, and so on.'[32] In more permanent camps, quarters of mud and thatch were built and the perimeter demarcated by a low mud wall or cactus hedge. A Nupe camp which Clapperton also visited was 'built of small bee-hive-like huts, thatched with straw, having four large broad streets' and a square near the emir's quarters. Apart from the armed men and beating of drums, it seemed no different from any ordinary large village. 'Here are to be seen weavers, taylors, women spinning cotton, others reeling off, some selling foo-foo and accassons, other crying yams and paste, little markets at every green tree, holy men counting their beads, and dissolute slaves drinking *roa bum*.'[33]

The towns of the West African forest were as extensively walled as those of the savannah and, as noted above, both the massive and the breastwork types of town wall occurred. As in the savannah, there is a great need for detailed and comparative study of these fortifications. Almost nothing

can yet be said with confidence about their dates. The difficulties in this connexion are illustrated by the case of the
ancient Yoruba city of Ile Ife, where the present main town
wall is known to have been built in the nineteenth century
but where there are remains of an extensive and far older
system of fortification about which tradition is silent.[34] In
the light of work so far accomplished,[35] however, two tentative generalizations may be essayed. First, following
Connah's suggestions mentioned earlier in this chapter, the
town walls in the forest seem almost always to have been
earthen ramparts, the use of free-standing walls of mud
bricks having been confined to gate-houses, the walls in their
immediate vicinity, and forts. Secondly, the additions of
a second and third wall, and in general the construction
of a more complex system of walling, were more prevalent
in the forest than in the savannah.

Only one detailed archaeological study seems yet to have
been attempted of any town wall in the forest, that by
Connah of the Benin walls. After citing the references to
these walls by the Portuguese visitor Pereira about 1500
and by the Dutchman 'DR' about 1600, Connah discusses
the strongly-held tradition that the innermost, and probably most massive, wall was built by Oba Ewuare, who
apparently reigned in Benin in the late fifteenth or early
sixteenth century. He then describes the complex pattern
of earthworks forming at least twenty separate enclaves
which his survey revealed. He argues that this pattern does
'not as a whole comprise a defensive system ... It is perhaps
rather the result of many different events spread over a long
time – an earthen record of a long process of fusion of
semi-dispersed communities which by the time of Ewuare
had grown into an urban society requiring defence at an
urban level.' Further research (which in the event has not
yet been undertaken) 'will perhaps confirm a suspicion that
a series of separate enclosures has gradually become linked

up to form a pattern within which Ewuare developed his fortified boundary.'[36] Comparison of the Benin walls with forest walls elsewhere suggests that they were not typical. The walls of Yoruba towns, for example, follow a course clearly related to the centre of the town, even if this course cannot be reconciled with what would seem to be tactical requirements.

The camps of the forest gave an equal impression of permanence to those of the savannah. The Dahomean army, for instance, made itself thoroughly at home when in the field. Snelgrave, visiting the king's camp in 1727, observed that: 'The Army lay in tents, which, according to the Negro-Custom, were made of small Boughs of Trees, and covered with Thatch, very much resembling Bee-Hives, but each big enough to hold ten or twelve Soldiers.'[37] Captain Jones remarked of the Egba in 1861 that they 'go to war in a very systematic manner as regards their present comforts'.[38] The warfare of the Yoruba in the nineteenth century, characterized by long sieges and generally semi-static conditions, culminated in the formation by the Ibadan and the Ekitiparapo respectively of two great opposing camps in Northeast Yorubaland, which were occupied from 1880 until the end of the Kiriji[39] or Sixteen Years War in 1893. Both these camps had developed by 1886 into regular walled towns, each containing from 40,000 to 60,000 inhabitants. The introduction of breech-loading rifles led to a great increase in the strength of the fortifications; the rude living quarters in the Ibadan camp were gradually converted into huts with thick mud walls which gave better protection against both weather and missiles.[40]

Most of the fortified towns of West Africa lie in level ground, though sometimes enclosing a hill which (as at Kano) was probably the original nucleus of the settlement and continued to be thought of as a possible last line of defence. There were also many hilltop towns and villages, though most of these have now been abandoned, leaving only fragmentary reminders of the occupation behind. Examples of such settlements abound in the rocky country of the upper Ogun river in Northern Yorubaland, a transitional zone between the forest and the savannah.[41] Some were occupied, or perhaps re-occupied, in the last century as citadels of refuge against marauding Fulani or Dahomeans, while others seem to be of earlier origin. These towns were usually encircled at or near the foot of the hill by a wall; this was sometimes built of stones where this material was easily come by. But the towns depended mainly for defence on the difficulty which an enemy would meet in ascending the steep slopes of the hill, the defenders of which dislodged boulders down the precipitous tracks. Many of the hilltops were provided with springs of water, but their confined space was insufficient to feed a population for long, and their great weakness was the inability to hold out against a protracted siege.

Towns lying beside rivers and lagoons had a special problem of defence since they needed to anticipate an attack by water. Turney-High has written that he knows of no water defences among non-literate peoples,[42] yet West Africa provides many examples of these. Where the terrain was suitable, a wall was built along the bank of the river, as at Abeokuta on the Ogun and at Kusseri on the Chari. The walls of the latter were described by Denham as being pierced on the riverside by two water gates.[43] Elsewhere, palisades were erected, the stakes extending into the water and backed by trenches and gun emplacements, as at Lagos and Epe during the English attacks in the mid-nineteenth century.[44] Robertson was told that Badagry had been defended in the eighteenth century by a forest of bamboo stakes which were so placed as to render useless the cannons mounted on the canoes of the Lagosian besiegers – who nevertheless succeeded in taking and destroying the town.[45]

Although the dry season brings a marked and rapid drop in the level of West African rivers, water defences provided a potentially formidable obstacle to an invader in pre-colonial times. The innovation of Idris Aloma of Bornu in building large boats to facilitate river crossings by his armies has already been cited (p. 125 above). Some two and a half centuries later Denham described the crossing of the Chari by a Bornu army advancing against the Bagirmi. 'The infantry, placing their spears and bags of corn on their heads, in their shields, cross with ease: the cavalry move over in canoes, and the horses swam (*sic*) at the sterns.'[46] Troops occupying defensive positions, on the other hand, seem to have had little appreciation of the value of water as an obstacle to an enemy. The cases of the Akwamu who moved across the Volta after their defeat by Akyem in 1730 and the Ijebu Ode who in 1892 based their defence against the British expedition on the insignificant river Yemoji[47] are exceptional. Probably more typical was the behaviour of the Whydah who, when invaded by Dahomey in 1727, made no effort to dispute the crossing of the lagoon which ran immediately to the north of their capital. '... the Pass of the River was of that Nature', Snelgrave writes, 'it might have been defended against (the) whole Army, by Five hundred resolute Men', but the Whydah set no guard there and 'only went Morning and Evening to the River side, to *make Fetiche* ... to their principal God, which was a particular harmless Snake ...'.[48] After their defeat, however, many of the Whydah did succeed in escaping with their king to the Popo islands where the Dahomeans, lacking boats and the knowledge of their use, were unable to follow.

As the prominence of fortifications implies, siege operations played a large part in West African warfare. The advantage in these operations lay on the whole with the defenders whose walls almost always enclosed a large area of farmland with sources of water, whereas the enemy, encamped in the dry season at a distance from their own farms and with a generally inadequate commissariat, soon felt the effects of thirst, hunger and other shortages. But the walls of the defenders were far too long to be fully manned, and there was always a danger of a sudden assault. Thus it was necessary to organize a system of sentries and alarms. Denham, that indefatigable and militarily experienced observer, who was himself consulted by Sheikh el Kanemi of Bornu about the manner of attacking a walled town, has described the constant patrolling of the walls and towers of Willighi. The same precautions were required in the camp of the besiegers, and Denham recounts how during the expedition against the Musgo in 1827 the Bornu commander formed a chain of Kanembu pickets on the side nearest the enemy, each consisting of five or six men, extending to two miles from the camp, the sentries calling to each other every half hour or so throughout the night. The alarm was given by striking two shields together.[49] Elsewhere, drums provided the alarm. Barth similarly noted in the Bornu camp 'a living wall of light Kanembu spearmen' keeping watch; he was less impressed by the guards of his personal camp whom on one occasion he disarmed while they were sleeping.[50]

Neither attackers nor defenders were content to remain forever passive. Though no evidence has been found of armed infantry patrolling, this may be assumed to have taken place; light cavalry were certainly used in patrols. From time to time the defending army would leave the shelter of their walls and engage the enemy in pitched battle, as happened at Ijaye.[51] A simple ruse was adopted by king Kpengla of Dahomey during his siege of the Yoruba capital of Ketu. Collecting his forces together, he withdrew as though abandoning the siege, thereby tempting the Ketu to move out from behind their great double walls and moat into the pursuit. The Dahomeans then turned on the Ketu and took some 2,000 prisoners, most of whom, Dalzel says, were later executed in Dahomey.[52] Attempts were also made to storm the walls of towns, as by the Dahomeans at Abeokuta in 1851 and 1864. Despite the presence of quite numerous cannon in West Africa, little use seems to have been made of artillery in such operations, though it is recorded that at Ijaye a number of piece of $1\frac{1}{2}$ inch calibre were used to some effect and the Dahomeans brought three six-pounders with them to Abeokuta in 1864.[53] Nor are there accounts of any use of siege engines, apart from the 'platforms' on which Idris Aloma mounted his gunners at the siege of Amsaka. Among the Mende the warrior entitled the Hakahouna, 'holder of the ladder',[54] must have been a key figure in scaling operations, while the incendiary arrows described in the last chapter were a useful aid to a force attacking a town.

The earliest accounts of siege warfare in West Africa are from the savannah. According to the *Kano Chronicle*,[55] a ruler of that town, Sarki Muhhamad Kisoki (1509–65), annexed the town of Nguru on the borders of Bornu, and in revenge a Bornu army marched on Kano. The defences were, however, too strong to be assaulted, and the defenders were presumably adequately supplied with food, so that the Bornu were eventually forced to retire. Much more detail is afforded in Ibn Fartua's narrative of the capture by Bornu, rather later in the sixteenth century, of Amsaka, a city of the So which was fortified by a stockade – presumably of sharpened stakes – surrounded by a deep ditch, and which had resisted all attempts at conquest by Idris Aloma's predecessors. After an abortive reconnaissance in force, the Mai waited some years – until about 1575 – and then set out in the month of Sha'ban preceding Ramadan. The army reached Amsaka about nine in the morning.

When the people of the town saw the dust of the Muslims rising to the sky, they mounted on to the tops of their houses and the pinnacles of the walls to observe what the Muslims would do. They were misled by the result of the former fierce engagement and the strength of their stockade. It is said that in their pride and self-sufficiency they said to the Muslims: 'You are as you were before, and we are as we were at first – and neither side will change – and none save the birds will see the inside of our stockade and town'. Thus they spoke to our people. Then came the Amir ul Muminin to their stockade to attack them in force. The enemy mounted above the stockade, and fired arrows and darts like heavy rain. No Muslim could pause a moment in the vicinity of the stockade without being pelted with arrows and hard stones which broke a man's head if they hit it. The stockade was full of people. The Sultan ordered the Muslims to fill in the trench which encircled the stockade with stalks of guinea corn, which the pagans had planted for food. They tried this plan for two or three days, but whenever the Muslims returned to their quarters in the evening, the enemy came out and took out of the trench all the stalks that had been put in it, so that nothing was left. This went on for some time. The Sultan then commanded the whole army to move and encamp close to the stockade to the north. They did so. This fighting took place in the month of God Sha'aban. The Muslims came to try and fill up the trench of the stockade in the early morning after they had shifted camp close to the town. Horns were blown and flutes sounded. There was all manner of noise in the stockade. The pagans set themselves to attack the Muslims in every possible way, and to divert them from the ditch by all kinds of wiles and devices; firstly by setting fire to thatched roofs, and throwing them down – a most formidable device; by poisoned arrows; by pots of boiling ordure; or by throwing hard clay which would split or break a man's head; then by throwing short spears, or

finally by throwing the long spear which is carried by
warriors. All these were among the methods of fighting.
They never ceased day or night. One day a section of the
stockade, about the length of a spear, was broken down.
The Muslims though they would gain entrance through
it and follow on. But the pagans built up the place with
mud in the open, the Muslims looking on and unable to
prevent them building. The Sultan then ordered the army
to cut tall trees to make platforms on three sides of the
stockade, so that the gunmen could mount on them and
easily shoot at the enemy inside the town in every direction possible.

They did as they were ordered. The people then worked
with a will in filling the ditch of the stockade with earth.
The place became flat and even. Our men began destroying and breaking up the stockade itself with matchets and
axes until they had cleared a large amount of the stockade
away, and so the enemy were hemmed in. To the Amir
ul Muminin and Caliph of the Lord of the Worlds, Al
Hajj Idris ibn'Ali (may God most exalted ennoble him
in both worlds) belong the credit for an apt device and
clever plan by which to fight the enemy – to wit, his order
to the army of the Muslims that the gunmen should get
in the first discharge so that they should not be fore-
stalled. Thus the enemy's hands would be rendered empty,
and victory over them would be easy. Thus it fell out.
The pagans began to shoot at the Muslims with showers
of arrows, but our army picked up all their shafts and
took them to the Amir in great numbers. It was impossible
to count the number of arrows collected. God alone
knows. Finally the arrows of the enemy came to an end.
Nothing was left. The heathen, therefore, assembled all
their blacksmiths, and asked their help in making arrows.
The Muslims outside the stockade heard the noise of the
beating of hammers on the big anvils which are placed

in the ground [i.e. with a stump fixed in the ground] for the making of new arrows. The enemy then began shooting with new arrows, covered with mud in place of the former poison, just to frighten the Muslims, pretending they had poison on them though it was not the case. The Muslims continued to destroy the stockade with matchets and axes, and filled the ditch, except that they left on the west side a part which they were unable to fill. After a time they destroyed the eastern part of the stockade, leaving a part in the middle. The enemy then became afraid, and lost their heads in terror and fright.

The day wore on. It grew hot. Then evening came. It was a Saturday [4th December, 1575 A.D.], the last day of Sha'aban. After sunset the new moon of Ramadan came out and the people saw it. The Sultan prayed the evening prayer, and refused to leave the battle. He ordered his tent to be brought, and slept among his troops. He directed the big drum to be beaten in a changing rhythm, to put fear into the hearts of the pagans. God cast into their hearts exceeding terror and a great dread. So they remained in fear and trembling, and ran from their stockade under cover of darkness. The Sultan followed them, killing the men and taking alive the women and children, tracking them down and following far without let or delay, so that none escaped among the heathen save a few. The people of every quarter came to him with many gifts, bowing their heads in submission, whoever they were and wherever they dwelt.

So there was destroyed this stockade which had held out against Idris' ibn 'Ali's predecessors. He turned upon it, and rendered it barren and deserted.[56]

A great slaughter of the captives followed, and then 'The Muslims all returned in the belly of the night, safe, with booty, joyful at the enemy's discomforture', except for one man who 'became a martyr in the confusion at night'.

Hunwick interprets the Bornu success at Amsaka as being mainly due to the 'psychological effect' of their newly acquired firearms, writing that Idris 'needed no other force than his musketeers'.[57] This conclusion does not seem to be sustained by the above passage which suggests rather that the victory can largely be credited to the infantry who showed courage and persistence in filling in the ditch and then attacking the stockade with their axes, qualities which reflected the discipline and leadership of the army as a whole. But it seems unlikely that the victory was obtained, as Ibn Fartua claims, at the cost of only one casualty.

An interesting comparison with Idris Aloma's attack on Amsaka is afforded by that of Sultan (or Caliph) Muhamad Bello of Sokoto on 'Coonia', a walled town of Gobir, in October 1826, of which Clapperton gives the following, rather jaundiced eye-witness account:

At 4 a.m. all was ready for commencing the war; but it was six before they started; the intermediate time being spent mostly in praying. I kept close to the Gadado, as it was his wish I should do so. Our path was through the plantations of millet and dourra of the enemy. At 8, the sultan halted under a tree, and gave orders for a camp to be formed, which was speedily done by the troops, cutting or pulling up the millet and dourra, and making huts, fences, and screens of the stalks. I waited on the sultan, who was dismounted, and sitting under the shade of the tree, near which he had halted. He was surrounded by the governors of the different provinces, who were all, with the exception of the governor of Adamawa, better dressed than himself.

After the midday prayer, all, except the eunuchs, camel drivers, and such other servants as were of use only to prevent theft, whether mounted or on foot, marched towards the object of attack; and soon arrived before the walls of the city. I also accompanied them, and took up my station close to the Gadado. The march had been the most disorderly that can be imagined; horse and foot intermingling in the greatest confusion, all rushing to get forward; sometimes the followers of one chief tumbling amongst those of another, when swords were half unsheathed, but all ended in making a face, or putting on a threatening aspect. We soon arrived before Coonia, the capital of the rebels of Goobur, which was not above half a mile in diameter, being nearly circular, and built on the bank of one of the branches of the river, or lakes, which I have mentioned. Each chief, as he came up, took his station, which, I suppose, had previously been assigned to him. The number of fighting men brought before the town could not, I think, be less than fifty or sixty thousand, horse and foot, of which the foot amounted to more than nine-tenths. For the depth of two hundred yards, all round the walls was a dense circle of men and horses. The horse kept out of bow-shot, while the foot went up as they felt courage or inclination, and kept up a straggling fire with about thirty muskets, and the shooting of arrows. In front of the sultan, the Zegzeg troops had one French fusil: the Kano forces had forty-one muskets. These fellows, whenever they fired their pieces, ran out of bowshot to load; all of them were slaves; not a single Fellata had a musket. The enemy kept up a sure and slow fight, seldom throwing away their arrows until they saw an opportunity of letting fly with effect. Now and then a single horse [*sic*] would gallop up to the ditch, and brandish his spear, the rider taking care to cover himself with his large leathern shield, and return as fast as he went, generally calling out lustily, when he got among his own party, 'Shields to the wall!' 'You people of the

Gadado, or Atego,' &c., 'why don't you hasten to the wall?' To which some voices would call out, 'Oh! you have a good large shield to cover you!' The cry of 'Shields to the wall' was constantly heard from the several chiefs to their troops; but they disregarded the call, and neither chiefs nor vassals moved from the spot. At length the men in quilted armour went up 'per order.' They certainly cut not a bad figure at a distance, as their helmets were ornamented with black and white ostrich feathers, and the sides of the helmets with pieces of tin, which glittered in the sun, their long quilted cloaks of gaudy colours reaching over part of the horses' tails, and hanging over the flanks. On the neck, even the horse's armour was notched, or vandyked, to look like a mane; on his forehead and over his nose was a brass or tin plate, as also a semicircular piece on each side. The rider was armed with a large spear; and he had to be assisted to mount his horse, as his quilted cloak was too heavy; it required two men to lift him on; and there were six of them belonging to each governor, and six to the sultan. I at first thought the foot would take advantage of going under cover of these unwieldy machines; but no, they went alone, as fast as the poor horses could bear them, which was but a slow pace. They had one musket in Coonia, and it did wonderful execution, for it brought down the Van of the quilted men, who fell from his horse like a sack of corn thrown from a horse's back at a miller's door; but both horse and man were brought off by two or three footmen. He had got two balls through his breast; one went through his body and both sides of the tobe; the other went through and lodged in the quilted armour opposite the shoulders.

The cry of 'Allahu Akber,' or, 'God is great,' was resounded through the whole army every quarter of an hour at least (this is the war-cry of the Fellatas); but neither this, nor 'Shields to the wall,' nor 'Why don't the Gadado's people go up,' had any effect, except to produce a scuffle among themselves, when the chiefs would have to ride up and part their followers, who, instead of fighting against the enemy, were more likely to fight with one another. There were three Arabs of Ghadamis in the army, armed at all points. Hameda, the sultan's merchant, was one. He was mounted on a fine black Tuarick horse, armed with a spear and shield, an Arab musket, brace of pistols, blunderbuss, sword and dagger. The other two, Abdelkrim, and Beni Omar, armed with musket, pistols, sword, and dagger. Abdelkrim was mounted; Omar on foot, who received a ball from the Coonia musket, which carried away his cartouche box, with all ammunition, early in the attack. The other two, Hameda and Abdelkrim, kept behind the sultan and Gadado the whole of the action, and always joined lustily in the cry of 'Allah Akber.' Once Hameda asked me, when I was near him, why I did not join in the cry: was it not a good place? I told him to hold his peace for a fool: my God understood English as well as Arabic.

The most useful, and brave as any of us, was an old female slave of the sultan's, a native of Zamfra, five of whose former governors she said she had nursed. . . . The heat and dust made thirst almost intolerable. Numbers went into the shade as they got tired, and also to drink at the river. When it drew near sunset the sultan dismounted, and his shield was held over him for a shade. In this way we continued until sunset, when the sultan mounted. We left the walls of Coonia for the camp. Upon the whole, it was as poor a fight as can possibly be imagined; and, though the doctrine of predestination is professed by Mahomedans, in no one instance have I seen them act as men believing such as doctrine. The feudal forces are most contemptible; ever more ready to fight with one another than they are with the enemy of their king and country, and rarely acting in concert.

The following day, 17 October, the attack was abruptly called off owing to 'the desertion of the Zurmie forces, and all the foot'.[58]

These two first-hand accounts of static warfare (in the sense of attacks on fixed defences) in the savannah may be usefully set alongside three accounts, mainly derived from European missionaries, of operations in the Yoruba forests: the two Dahomean assaults on the walls of Abeokuta in the mid-nineteenth century, and the long ordeal of Ijaye under siege by the Ibadan from March 1860 to March 1862. At Abeokuta on 3 March 1851 the Dahomeans, 'all armed with guns', made ten full-scale attempts to storm the walls, along which the Egba, supplied with ammunition by their women in the rear, fought back with their muskets and cutlasses. By early evening the attackers, having suffered some 2,000 casualties, realized that the day was lost and broke off the engagement. Despite harassment by the Egba, they succeeded in making an orderly withdrawal behind their borders.[59] Thirteen years later they came again, determined on their revenge. Setting out in February 1864, towards the end of the dry season, the army of some 10 to 12,000 marched in four 'battalions', bringing with them three brass six-pounder guns mounted on locally-made carriages. There are conflicting reports as to whether king Gelele accompanied them. After taking twenty-two days over their march of some 120 miles, they reached Abeokuta at dawn on 15th March and crossed the river Ogun under cover of mist. Meanwhile, the Egba, after hastily repairing their neglected walls and cleaning out the ditches, had been manning their fortifications through the night. On having the alarm given by their cannon at the Aro gate, they sent out a detachment of 400 warriors but this, failing to draw the Dahomean fire, soon returned to the town through tunnels (so Burton says) under the walls. The Dahomean army then deployed into three groups, forming a line some 700 yards long, its right hand group opposite the Aro gate. These now advanced towards the wall, one column approaching to within 200 feet and the rest remaining at a distance of 300–400 feet. At about 7 a.m. they opened fire, and under cover of the smoke from the muskets a party of Dahomeans entered the ditch; of these a few – mostly Amazons – succeeded in scaling the wall, but were at once slain. During the course of the next hour some seventy to eighty Dahomeans were killed in the ditch. The Dahomean army then retired from the walls, but their retreat, at first orderly, quickly became a rout as the Egba left their defences in pursuit. The total Dahomean losses were estimated at over 2,000 dead and 2,000 taken captive (of whom only one in four was said to have been born in Dahomey). They had also lost their brass guns – of which they seem to have made little or no use, as they are otherwise unmentioned in any account – and much ammunition and equipment. The Egba had lost only 40 men, with 100 men wounded.[60]

During the siege of Ijaye,[61] there seem to have been no attempts to storm the walls of the town, operations taking the form of a not very close investment with occasional pitched battles fought in the fairly open country beyond the walls and smaller-scale engagements elsewhere within the area of Ijaye's former predominance. The Egba army, sent to Ijaye from Abeokuta, settled into a walled camp abutting on the south-west walls of the town, while the Ibadan encamped at Olorishaoko, some seven miles to the south. The capture by Ibadan of the hill-town and fortress of Iwawun, key to the vital food supply of the Oke Ogun, in December 1860, the death of Kurunmi, the Ijaye leader, in June or July 1861, and the threat of a renewed Daho-

mean attack on Abeokuta, all combined to lower the resistance of the Ijaye, already suffering many privations. Finally,
the departure of the Christian missionaries resident in the
town and the almost simultaneous desertion of the Egba
brought about the sudden collapse of the defences in March
1862.

The naval equivalent of siege warfare is the blockade. An
example of this occurred during the war of 1783 between
Dahomey and Badagry. The Dahomeans had obtained the
assistance 'by bribes or promises' of the ruler of Lagos
who accordingly stationed a fleet of thirty-two large canoes
in the lagoon to the east of Badagry, thereby cutting off
that town, whose narrow territory of sand and swamp
yielded little farm produce, from their usual source of food
supplies. At last the people of Badagry made a desperate
bid to break out of the town in their canoes, in the course
of which they lost many killed and captured.[62]

The importance attributed in West Africa to fixed defences
and the predominance of forms of warfare revolving around
these defences suggest that West African wars should provide examples to test the first part of Andrzejewski's theory
that, other things being equal, a defensive attitude to war
promotes the territorial division of political power, while
conversely an offensive posture tends to promote political
concentration; where defensive considerations prevail, the
capital of a state will encounter rival political centres, and
each town will be able to control only a limited area.[63] This
theory is to some extent supported by the history of Yorubaland after the breakdown of Oyo in the early nineteenth
century. The Alafin had been forced onto the defensive by
political developments to the north and to the south of their
kingdom, and Atiba's device of decentralizing responsibility
for dealing with external threats to security (p. 159 below)
amounted to little more than a recognition of the situation

already created by the founders of Ibadan and Ijaye; power was now dispersed not only to these two towns but also to a host of lesser 'places of refuge' within or near to the forest belt. After the Ijaye war, Ibadan set out to create a new 'empire', extending in some areas beyond the confines of the old Oyo one, which involved an attempt to recon-centrate political power by binding the subordinate towns even more closely to Ibadan than they had been bound to Oyo and previous overlords. Meanwhile, far to the north, the medium-sized Habe kingdoms had been swept away by the Fulani jihad in its first aggressive phase, the walled cities of the Hausa being conquered as much by Muslim sympathi-zers within as by the warriors without; on their political ruin was established the empire known as the Sokoto Caliphate. Yet Fulani aggression, so far from providing an example of political concentration, resulted in the formation of a loose confederation of largely autonomous emirates.

The conclusion must be that 'other things' are rarely, if ever, 'equal' in the real world. Andrzejewski's maxim remains a useful guide to the interaction between military and political theory and the practical results of such inter-action. But there will always be many other considerations to be taken into account in any discussion of the impact of war and preparation for war on a country's internal situation. These include the nature of the leadership and the ideology of the forces engaged in any struggle for power, the character of the terrain on which they operate, and the types of armies and armaments in the field. All these are of at least equal importance to, and often of greater im-portance than, the strategy and tactical aims of the power-seekers, whether at peace or in war.[64]

VII. Strategy, tactics and the battle

The conduct of operations in war may conveniently be treated under the twin aspects of strategy and tactics. These correspond broadly to the scale of the undertaking being considered and, though they become blurred at one end of this scale, can usually be clearly enough differentiated in societies which have reached the level of political organization labelled 'statehood'. Strategy, the art of war in its highest manifestation, is but a step from the policy pursued by a government in its international relations, or is indeed a part of that policy; it is Clausewitz's 'continuation of political intercourse, with a mixture of other means'.[1] Tactics, the conduct of war in the field, corresponds to the work in peaceful conditions of the diplomatist as he carries out the directions of his government. In strategy, the objectives are long-term, and arrangements for their achievement must be based on an assessment of the relative powers, including the will-power, of the protagonist himself and of his enemy. Tactics, associated with the leadership of troops in contact with an enemy in a campaign or battle, have shorter-term ends – the destruction of a specific group of the armed enemy, the gaining of a tract of land or a fortress, for example – and require equally careful and original planning, local intelligence, and ability in the use and combination of weapons.

A consideration of past military strategy must begin with an attempt to understand the plans and ambitions which precede it. History in such terms presents much difficulty since the motives of governments as well as of men are confused and changing, and often lie far below the surface of consciousness. The difficulty is increased further in the case of the history of non-literate peoples who have left behind few if any indications about their policy and the motives for their actions to supplement the possibly misleading record of 'what actually happened'. Earlier in this book (in chapter III), the general aims of West African wars were examined, aims which like those of wars elsewhere can be summed up under such headings as booty and wealth, trade, territorial expansion, the desire for power over others, and ideology. To achieve such aims, long-range planning and preparation, political and diplomatic as well as military, were needed, culminating in the launching of operations at a time and place and in a degree and combination of strength all previously determined upon as necessary ingredients in the receipt for success. Only a part, usually a very small part indeed, of this can be confidently identified in the case of the warfare of pre-colonial West Africa.

Once again the account by Ibn Fartua of Mai Idris Aloma's wars provides the earliest evidence. Perhaps the most striking of the Mai's campaigns was one only indirectly referred to by Ibn Fartua, that to the Kauwar oasis far to the north of Bornu which, as already indicated, may have been intended to secure long-distance trade and to keep open Bornu's links with the distant Mediterranean seaboard. If this was so – and there is no direct evidence to support this speculation – then here was strategy on a grand scale. Rather more can be discerned of Idris's wars in Hausaland and Kanem. The campaigns against the people of Kano, possibly originating as a response to Kano incursions into Bornu territory, as Ibn Fartua alleges, seem to have been first intended to set up a secure and defensible frontier in the west, a limited objective which success expanded into an ambition to dominate Hausaland itself.[2]

The five wars conducted against the Bulala in Kanem seem to have been actuated by the desire to dominate a people whose dynasty had a vague relationship with that of Bornu, the Sefawa, and on whose territory the Mai had some claims. After the conclusion of the treaty of Siki which demarcated the boundary between Bornu and Kanem, evidently in Bornu's favour, Ibn Fartua records that 'All our commanders and captains rejoiced at our increase of territory.' Finally, Idris Aloma's policy of denying access to their grazing lands of certain Tuareg groups who had raided his country is an interesting example of deliberate 'ecological dislocation', though one which should perhaps be classed as tactics rather than strategy.[3]

The introduction of firearms into West Africa affected the strategy of warfare even more profoundly than its tactics. Access to the coast, where arms and ammunition could be obtained direct from the European traders, became an objective of first importance for the more vigorous and ambitious of the inland kingdoms of the forest. As early as the reign of Opoku Ware (1720–1750), the Ashanti government, already the dominant power in the hinterland of the Gold Coast where its infantry had successfully challenged the cavalry of the savannah peoples, was sending exploratory messages about trade to the Europeans, and before mid-century the Fante who blocked their way were made aware of this Ashanti *Drang nach Suden*. For the next two reigns at Kumasi the Ashanti-Fante duel was as much diplomatic as military, although at least three invasions of Fante took place in the eighteenth century. With the accession of Osei Kwame as Asantehene in 1777 the drive to the sea was pursued with increasing vigour and led to the expeditions of 1806–7 and 1811.[4] The Fante, organized into numerous kingdoms as the Ashanti had been before the consolidation effected by Osei Tutu and his successors, found increasing difficulty in resisting these incursions and would probably have failed in the task had it not been for the help of the Europeans ensconced in their trading forts along the coast. In 1824 the Ashanti defeated a British-led force at Asamankow, but two years later, at Dodowa in 1826, they were themselves defeated by a combination of the Fante allied with Asin, Akim and Akwamu, states which had revolted from Ashanti domination, assisted by the Danes and under British command. Dupuis, writing before these two battles, praises the Ashanti for the Napoleonic boldness of their movements: 'The military policy of an African government, is conspicuous in the cabinet maxim of Ashantee, during the war with Fantee. It may in some degree be assimilated to the modern tactics of recent introduction to Europe, by which the armies of France, under their late leader, pursued a direct object, however remote, with that celerity which frequently guaranteed its own success, and leaving half the war behind, terminated the campaign by a single blow at head quarters.'[5] The comparison may seem far-fetched, but the boldness of the Ashanti command's strategic concept and the skill and persistence with which this was implemented were certainly outstanding in West Africa.

The objectives of the Dahomean drive to the sea under king Agaja in the first part of the eighteenth century have been variously interpreted. John Atkins, writing eight years after the overrunning of Whydah, thought it probable that the king was 'incited to the Conquest from the generous Motive of redeeming his own, and the neighbouring Country People from those cruel Wars, and Slavery, that was continually imposed on them' by the King of Allada and by fetish rule in Whydah, while at the end of the century Dalzel, more cynically, considered that it was the evident advantages of the firearms obtainable there which made the king 'determine to possess himself of a part of the coast . . .'.[6] Subsequent writers have expanded Dalzel's explanation to include a general desire to participate directly in the lucrative slave trade. Akinjogbin has questioned this thesis, arguing that Agaja had no difficulty in obtaining firearms and that his motives were rather to replace the 'traditional political system' of the coastal Aja, which no longer provided orderly conditions, and 'to stop by stages the slave trade in the Aja country and substitute for it a general trade in agriculture'.[7] This revision has been challenged by Law,[8] and it is difficult to believe that the prospect of direct access to the arms traders did not weigh with the king. But whatever his motives – and the evidence is insufficient to reach any firm conclusion here – Agaja prepared his invasion with great skill and deliberation, instituting his intelligence service, training a disciplined army, and sending his representatives to the coast. His conquest of Allada in 1724, giving him control of two ports, Jakin and Offra, was followed by that of Whydah in 1727. These conquests, however, involved him in a war with Oyo whose forces invaded Dahomey in 1728, 1729 and again in 1730, until under the treaty of 1730 Dahomey, while agreeing to pay tribute to Oyo, was allowed by Oyo to retain possession of Whydah and a large part of the territory of Allada.

The expansion of Oyo to the coast some 200 miles southwest of the capital took place probably in the seventeenth century, but owing to the absence of European observers on the scene, direct evidence for this early period is lacking. The expansion was effected through the 'Dahomey gap' in the forest, where the persistence of savannah-like conditions enabled the Oyo cavalry to establish the influence of the kingdom on the coast, first at Allada and Whydah, and later at Porto Novo and then Badagry. Oyo thereafter became a participant in the Atlantic slave trade, a development which increased both the wealth and the political problems of the kingdom.[9] The maintenance of free movement through the trade 'corridor' which had been created (running through Egbado territory) provided a problem in strategy, solved by the consolidation of Oyo power within a series of towns, some of which were new foundations by immigrants from Oyo. Though there is controversy as to the period at which these towns were founded,[10] it is agreed that the ruler of Ilaro, the Olu, represented the Alafin among the Egbado as a whole. The good order which was kept in the area was vividly illustrated by the safety and ease with which Clapperton and Lander travelled up to Oyo by this route in 1825–6. Yet by this time the Oyo kingdom was already beset by serious internal troubles allied to the aggression of the Hausa-Fulani now established in Yoruba country at Ilorin. The situation was exacerbated by the raids which the Dahomeans, after refusing their annual tribute to Oyo in the 1820s, were making into Egba and Egbado country, menacing the vital economic interests of Oyo and the security of all Yorubaland. After the fall of Old Oyo and the re-establishment of the kingdom on a much diminished scale in the forest to the south, Alafin Atiba attempted a final piece of grand strategy in the manner of his forebears by devolving on Ibadan the duty of protecting Yorubaland against the Ilorin to the north and north-east and on Ijaye that of guarding the west against the Dahomeans and other enemies.[11] But, as already pointed out, this hardly amounted to more than a recognition of the existing division of power, a division which in the event soon broke down amid the internal rivalries of the Yoruba.

The holy wars fought by Muslim zealots in West Africa have been cited already as instances of wars originating in and fought for motives which purported to be, and probably quite largely were, ideological. They also illustrate the capacity of West Africans for strategical planning and action. The most prominent of the jihads, that led by Sheikh Usman dan Fodio and his sons in Hausaland, was based on a series of local uprisings which enabled the leaders to gain control from within of the strongly walled towns which formed the centres of power in the various Habe and other kingdoms. To stimulate, aid and lead these risings, Usman dan Fodio sent out lieutenants, commissioning each with a flag as symbol of his office and authority. Thus, as Smaldone remarks, the jihad had the advantage of operating on interior lines, and as a series of almost simultaneous risings were mounted at numerous points, the Habe rulers were prevented from any effective combination against the jihadists.[12] This series of independent campaigns, beginning in 1804, had led by 1808 to the emergence of a number of separate emirates under the flag-bearers. These were virtually coterminous in territory with the Habe and other kingdoms which they had overthrown and were ultimately all subject to the authority of the Caliphate set up in the new town (and former camp) known as Sokoto (though the western emirates were subject in the first instance to Gwandu and only indirectly to Sokoto). Thus the Caliphate was a conquest state administered according to a highly decentralized plan, its constituent emirates each having a good deal of autonomy. The work of extending the frontiers of dar al-Islam continued for much of the rest of the century, but warfare against pagans and slave-raiding did nothing to improve and little to maintain the military qualities of the empire, and in the third year of the twentieth century this 'state organized for war' collapsed after only a feeble resistance to the advancing British.

To the formation of policy and its military implementation in strategy, the gathering of intelligence is an indispensable preliminary. It is equally necessary before the decisions are taken as to the conduct of battle, which constitute tactics. Nor does intelligence end with the opening of action since a commander must aim to pierce the fog of war by the use of secret agents and the sending out of mobile, lightly-armed troops to probe and observe. Evidence about the intelligence services of the armies of pre-colonial West Africa is naturally difficult to obtain, but it is certain that this aspect of warfare was far from neglected. The role of the famous Dahomean spies, the *agbadjigbeto*, is well-known from travellers' accounts[13] and reference was made in chapter II above to the part they played in Agaja's invasion of Whydah in 1727. Both Bowdich and Dupuis refer to the Ashanti 'corps of observation'; these were spies who moved by night in enemy country and by day lay hidden but alert in the branches of high trees.[14] According to Barth, many spies operated in the territory of the Sokoto Caliphate on behalf of the Habe kings of Gobir, while during a war in 1853 the Gobirawa apprehended and executed the chief spy of Sokoto in their capital.[15] More open reconnaissance was pre-eminently the duty of light cavalry and may be assumed to have been carried out in all armies in which there was a cavalry arm.[16] Observation of enemy movements was also obtained from the towers and look-out posts on the walls of towns and camps which were described in the last chapter. Those selected for duties of observation and reconnaissance must naturally have been men with keen eyesight and other well-developed physical senses. The only reference to the use of artificial aids to observation known to the writer is the report by Captain Jones that the Yoruba war chiefs at Ijaye in 1861 had 'either reconnoitering glasses or telescopes with the use of which they are quite familiar'; earlier in the century, however, Denham had explained the use of his telescope or 'spy glass' to the rulers of Katagum and Kano.[17]

When intelligence has played its part, a military commander is ready to decide on the method and timing of his approach to the objective. In West African warfare night attacks seem to have been rare; to take examples from three peoples living in different habitats and with different forms of fighting, they were specifically avoided by the Nupe, the Ashanti and the Kagoro (the last treated their small-scale wars very lightly, going out in the morning with packed luncheon and returning home punctually at nightfall). On one occasion Idris Aloma marched the Bornu army through the night to a pagan town and then waited until the enemy woke from sleep at sunrise before putting in his attack.[18] Movement in darkness followed by a dawn attack were the most commonly practised tactics in West Africa. It was adhered to by, for example, the Dahomeans, who sent out soldiers disguised as traders to kidnap enemy stragglers and then force them to act as guides through an unknown and hostile country in darkness.[19] The Jukun army was accustomed to set out either at dawn or in the evening, as the royal diviner directed, but always put in their attack at first light. Finally, high noon was sometimes chosen for a surprise attack by Ibo warriors as the defenders were likely to be absent then on their farms.[20]

The order in which West African armies made their approach march was dictated by the nature of the country. Along the forest paths, single file had of necessity to be kept, whereas in the open country of the savannah and the semi-savannah the more conventional column was the normal formation, with a more or less differentiated van, main body and rear guard. As already mentioned, the titles of the chiefs generally indicated their positions in battle and those of the troops under their command. In the Dahomean army the primary distinction was between chiefs of the right and of the left hand, and a similar system obtained among the Oyo and later the Ibadan.[21] The *asafo* of the Gold Coast was organized into companies of the right and left wings and the centre.[22] Presumably on the march these wings closed up into column of line. In Mai Idris Aloma's army the 'shield-bearers' (spearmen) preceded the cavalry. The Jukun moved in five groups: the heavy cavalry in the lead, with quilted covers on their chargers, the light cavalry, whose mounts were unprotected, followed by infantry with thrusting spears, infantry with throwing spears, and finally bow-men.[23] Wherever a body of mounted troops (as opposed to mounted chiefs or officers only) existed and the nature of the country allowed, a light screen of cavalry protected the front, rear and flanks of the whole force as it moved. The army of Fulani Zaria, for example, was preceded by mounted scouts and pathfinders and itself consisted of a main cavalry force followed by infantry and bowmen, the ruler with an escort of heavy cavalry, infantry reserves, and a mounted rearguard, with light cavalry operating on the flanks.[24] In the Hausa-Fulani armies of the mid-nineteenth century musketeers were at first kept in the centre but as they assumed greater importance, and increased in number, they were moved forward into the van. Smaldone has identified two basic tactical formations used by the armies of the Sokoto Caliphate when on the move towards an enemy. The first, that preferred by Muhammad Bello, consisted of the cavalry followed by infantry and then by the reserve. In the second formation, that used by the Nupe (and also by the Hausa of Abuja), the infantry were in the lead with the cavalry and then the reserve coming up behind.[25]

The loose discipline which to Europeans seemed characteristic of West African armies doubtless played havoc with these ideal arrangements. Clapperton's vivid account of the march of the Sokoto army against the Gobir town of 'Coonia' in 1826, cited in the last chapter, illustrates this. Dahomean armies were more orderly and thus more capable of being disposed and manoeuvred according to the commander's plans, as was shown in their attacks (albeit unsuccessful) on Abeokuta in 1851 and 1864. But confusion and muddle can beset the best trained of armies, as those with experience of war in the field testify. Moreover, much of the confusion described by Europeans was more apparent than real. Denham, for example, noted of the Bornu army in its expedition against the Munga that though it seemed to keep little order before it came into contact with the enemy, when battle approached everyone knew his place.[26]

'War is trickery', observed the Prophet Muhammad, and according to a more recent authority, Professor Falls, 'Surprise is the most effective of all keys to victory'. The military commanders of West Africa were well aware of the truth of these maxims. King Agaja of Dahomey, according to a tradition which Akinjogbin pronounces 'quite credible', preceded his attack on Whydah in 1727 by suborning his daughter, a wife of King Huffon of Whydah, into pouring water over the gunpowder in her husband's arsenal. Again, the Dahomeans deliberately spread false reports about the route to be followed by their armies and generally approached an objective circuitously, often doubling back on their tracks in order to take the enemy from an unexpected angle[27] – a manoeuvre which perhaps few other West African armies were capable of carrying out successfully. The need to prevent information reaching an enemy and his spies led commanders to keep secret their intentions until the last possible moment and to give misleading reports, as did the leader of the Bornu expedition which Denham accompanied against rebels in Kanem.[28]

On arrival before its objective, whether this was an army drawn up in open country or a town or camp sheltered by its walls, a West African attacking force deployed, usually into line.[29] Before the introduction of firearms in large numbers, battle was opened by the spearmen who advanced towards the enemy throwing their spears, while in the rear archers discharged their arrows over their heads or, as Clapperton writes of the Kano army, 'shooting from between the horsemen as occasion offers'.[30] Where cavalry was on the field, they were probably held back until the missile weapons had made their impact and were only then ordered to charge. When firearms came into general use, the musketeers led the way. Denham's observation that in

Bornu 'As in our quarter of the globe' it was the infantry who usually decided the battle[31] has already been cited. Yet Barth, writing some thirty years after Denham, saw the cavalry as the essential element in the armies both of the Bornu and the Hausa-Fulani states. He considered that 'whatever military strength the Kanuri may still possess, it is almost solely to be looked for in their cavalry', a judgement which probably reflects the mistrust which the Shehu El Kanemi felt for the Kanembu spearmen whose loyalty the displaced dynasty of Bornu had commanded.[32]

Contact between opposing armies was preceded and accompanied by a confused and appalling din set up by the beating of drums, the blare of trumpets, the shouting of battlecries, and the rattle of musketry. Noise was a weapon in the contest: Sheikh Usman encouraged the warriors of the jihad to go into battle declaiming their genealogies, reciting heroic verse and boasting of their exploits, and the loud reports of the guns were often more effective than their bullets in the alarm they caused to men and horses.[33] While the air darkened with the storm of arrows, javelins, and darts, 'as thick as a cloud of locusts' as Ibn Fartua puts it, the infantry moved forward, roughly in line if the ground permitted or bunched around their commanders – 'close together in a sort of crowd', Bosman observed – until warrior met warrior in a mêlée through which horsemen tried to force a way with their thrusting spears, and which lasted until one side or the other gave way. 'They fight in such a way', explained Labat, 'that once their lines are broken, there is no hope left; the choice is between precipitate flight and massacre.'[34]

For those with swords, thrusting spears, clubs and similar weapons, maintenance of the attack or holding of ground depended almost entirely upon leadership and morale. Those who wielded missiles had the added problem of replenishment. For the archer, it was a matter of seconds to take a fresh arrow from his quiver, re-span his bow, and fire – and possibly too he kept a few extra arrows stuck in the ground beside him, as did the English at Crecy[35] – while the throwing spearmen armed himself with a bundle of spears or darts. As the contest grew hotter, fresh supplies were needed, and these were generally brought forward (as Denham saw at Musfeia) by women and slaves, while the enemy missiles could also be turned to account, as they were by the Bornu warriors at Amsaka. The musketeer had greater difficulty, reloading being for him a complicated and even dangerous task. His rate of fire can rarely have exceeded one round in three minutes and was probably usually much slower. Accounts speak of musketeers with-drawing to a distance before re-loading. Bosman writes that the Fante soldiers in the late seventeenth century 'creep towards the Enemy, and being come close, let fly at them; after which they run away as fast as they can, and ... get to their own Army as soon as possible, in order to load their Arms and fall on again'. Two hundred years later the Yoruba, to whom the general use of firearms came late, were conducting their battles in a rather similar fashion, each side retiring at intervals to replenish its ammunition (though not necessarily between individual shots). Captain Jones describes the sedate and not very lethal operations at Ijaye in 1861 as 'a succession of advances and retreats throughout the day, until sunset when, ammunition ex-pended, begrimed and tired, each side draws off, victory being claimed according to the returns of killed and wounded, which usually is not large'.[36]

In pitched battle, as opposed to the manoeuvres preceding battle, only rudimentary tactical skill was exhibited. Apart from the initial barrage of arrows and musket fire, there was no concept of covering fire and movement, and once action began there was little possibility of varying the plan of battle according to circumstances. This was especially the case in forest fighting where the command could attempt to exercise control only by word of mouth or by drum or trumpet signals, all but impossible to distinguish in the noise and confusion. The basic stratagem of allowing a weak, or apparently weak, centre to fall back early in the action, following this by an attempt to envelop the pursuing enemy from one or both flanks, may have been practised (for example, by the Ashanti at Emperou, described below), but more popular were attempts to concentrate an attack on one or both wings of an enemy army. Manoeuvres of this kind, more practicable in the open conditions of the savannah than in the forest, in Turney-High's opinion distinguish 'primitive' from 'true' warfare.[37] A more forthright scheme of battle seems to have been preferred by Usman dan Fodio who advised the military commanders of the jihad always to place the bravest and best armed in the centre, for so long as the centre held, no harm could come to the army[38] – advice which was probably given in recollection of the decisive resistance of the centre of his army to the Gobirawa in the great battle of Tabkin Kwatto.

The main differences between warfare in the West African savannah and in the forest may now be apparent. The savannah was suitable for the use of horses on a scale and in a manner reminiscent of the cavalry wars of Europe and Asia, making possible the creation of mobile forces which could move swiftly to an objective and also as shock troops exert on an enemy pressure weightier than could be applied by any body of dismounted men, the horse being the equivalent both of the modern troop-carrier and of the tank. Yet wherever firearms had been introduced in any great numbers, their effect tended to diminish the impact of the heavy cavalry charge.[39] Light cavalry supplemented the heavy by protecting the flanks of an army by skirmishing and reconnoitering. Again, the relatively open country allowed infantry to move more easily and quickly than through the forest where the narrow, rutted paths twisted from side to side, every fallen tree causing a new diversion. (Even the well-disciplined, British-led expedition from Lagos to Ijebu Ode in 1892 could cover only eight miles in a day during their march from the lagoon-side.)[40] At the objective itself, an attacking army in the savannah could deploy speedily and without its commanders losing control over their men. Finally, long-range weapons, especially the bow and arrow, could be brought into effective use. Yet none of these advantages to savannah warfare should be taken as applying in an absolute sense. Physical conditions varied considerably as the savannah shaded on the north into the desert and on the south into the forest. Much of the West Sudan is covered by orchard bush and long grasses, providing a hindrance if not an obstacle to the movement of a body of men or horse. This characteristic seems to be the explanation of the frequent references by Ibn Fartua to the cutting of trees by the armies of Idris Aloma in their campaigns, the trees being those which formed protective thickets and hedges around the towns and villages of the enemy.[41]

Two battles fought in the savannah of which fairly full accounts have been preserved are those at Tabkin Kwatto (1804), when the army of the Habe state of Gobir was routed by the forces of the Fulani jihad, and at Angala (1824), where the Bornu army drove back an invasion from Bagirmi. The first, Tabkin Kwatto,[42] followed close on the flight – accounted his 'hijra' on the analogy of Prophet Muhammad's flight to Medina – of the Sheikh Usman dan Fodio from Degel to Gudu in the barren scrubland north-west of the present town of Sokoto. At Gudu the Sheikh's supporters quickly swelled into a force ever more dangerous to the security of the Gobir kingdom, while his assumption of the title of Amir al-Mu'minin threatened all the states in the region. Yunfa, the Sarkin (king) of Gobir, could not allow this situation to continue unchecked, and in June, just before the onset of the heavy rains, he took the field against the Muslim insurgents, threatening Gudu from a camp at Ayame some twenty miles to the south. After rounding up some Fulani cattle herds, he left his camp and made a wide sweep westwards, apparently with the aim of bringing the Fulani to battle and meanwhile cutting off their retreat to the west. Eventually he took up position with his army by the shore of a small lake called the Tabkin Kwatto, some twenty miles west of Gudu. The Fulani, who had only a few days before returned from a swift and successful action against the town of Birnin Konni (Clapperton's 'Coonia'?) and were apprised of the movements of the Gobirawa, made a night march to Tabkin Kwatto and, as the poem by Abdullahi, Usman's brother, recounts, sighted their enemy as they rested in camp, their 'spitted meats around the fires', their tents spread with carpets and cushions.

The Sarkin Gobir's army consisted of 100 heavy cavalry wearing chain mail and quilted arrow-proof armour ('Upon them were ample suits of armour, And beneath them fine long-necked horses', writes Abdullahi), a few Tuareg mercenaries, and an unknown number of infantry, probably amounting to several thousand, armed with bows and arrows, swords, spears, javelins, and battle axes, and carrying shields; there were also a few with firearms. The Fulani army, commanded by the Sheikh's son, Muhammad Bello (later Sultan of Sokoto and Caliph), and brother, Abdullahi (later Emir of Gwandu), was similarly equipped except in two important particulars: that they had no firearms and only some twenty horsemen, the latter being defectors from the Gobir cavalry. Both armies consisted basically of Hausa soldiery, including on the Fulani side a contingent of Zamfara (who had deserted from their alliance with the ruler of Gobir), with an admixture of Fulani and Tuareg. Apart from their superiority in cavalry and firearms, the Gobir army was also the more numerous.

On arrival at the lake soon after daybreak on 21 June 1804, the warriors of the jihad refreshed themselves, watered their horses, and performed their ablutions. Meanwhile the Gobirawa withdrew to the low hills north of the lake, where they formed line of battle. Perhaps, as Johnston considers, the Fulani would have been wiser to fight a defensive action. Instead, filled with zeal, they left the shelter of the trees around the lake and at about midday opened the attack. As they advanced, their banners flying, they looked to their enemy, says Abdullahi, 'like an ogre in striped clothing'. The Gobir infantry met them with volleys of musket fire, but this proved ineffective. Thenceforth, Johnston writes, 'like Agincourt which it strikingly resembles, the battle developed into a contest between the shock of heavy cavalry and the attrition of lightly armed but highly skilled archers'. Bello himself described the course of events:

> The enemy made ready and took up their positions. They had donned mail and quilted armour, and with their shields they formed their line against us. We too formed our line against them and every man looked squarely into the eyes of his foe. Then we shouted our battlecry three times, *Allah Akbar*, and charged against them. When the two lines met their right wing overbore our left wing and pressed it back upon our centre. Their left wing also overbore our right wing and pressed it back upon our centre. But our centre stood fast and when our right wing and our left wing came up against it they too stood fast and yielded no more. Then the two armies were locked together and the battle raged.[43]

This makes it clear that the numerical superiority of the Gobirawa enabled them to envelop both wings of the Fulani army which was then compressed into a square. Within this square the bowmen, presumably protected by the spears of the other infantry, managed to keep at bay the attacks of the Gobir cavalry, which seems to have been hampered by the narrowness of their position with the lake on their flank. At length the archers mastered the horsemen and the Gobirawa broke and fled, leaving their tents and much other booty to the victors.

The battle of Tabkin Kwatto was the first major engagement of the holy war of the 'Fulani. Their success greatly encouraged the revivalists and ensured the survival of their movement. Bello ascribed the victory to divine intervention: 'The Lord broke the army of the godless...', he wrote.[44] More dispassionate observers, after acknowledging the part played by morale in the battle, add to this reason the superiority of the Muslim archers and the square formation which the Fulani army was forced to adopt and which – as in the days of the pike in Western Europe – proved an effective disposition for defence by infantry against cavalry. 'The Muslim victories of the early *jihad* period', Smaldone concludes, 'represented a temporary triumph of firepower over shock power in the history of Hausaland.'

But Tabkin Kwatto was not a decisive victory. The following December the Sheikh's forces were defeated by an allied Habe army at Tsuntua, losing 2,000 dead, and again in the following year at Alwassa. But by now the Fulani were making increasing use of cavalry, both heavy and light, which they integrated with their infantry. The principal infantry weapon continued to be the bow, the archers being now subdivided like the cavalry into heavy, or armoured, and light groups. The support given by the archers to the cavalry provides one reason for the Moslem victories. But above all, in this first phase of the jihad, the Fulani were operating on interior lines, stimulating revolts in all the Hausa kingdoms and overcoming from within the strongly walled cities of the land.

Since firearms had been known in parts of the Western

Sudan at least since the sixteenth century, and their value had been demonstrated by the Moroccans at the battle of Tondibi which overthrew Songhay in 1591, it is at first surprising that they played so small a role in the battles of the Fulani jihad in the early nineteenth century. The point has already been discussed in chapter V, and it may be recollected that Smaldone attributes it mainly to restrictions on the trade across the Sahara, rather than to any Fulani fear of the social effects of their introduction, while the poor quality of the arms available may have been an even more important reason. But the Muslim leaders also seem to have regarded firearms with some distaste as unworthy of the upright followers of the Prophet and their use in the jihad was relegated to selected servants and slaves.[45] Later, when supplies increased in the decade between 1860 and 1870, standing forces of musketeers were formed.

In Bornu the situation was rather different. Denham noted that the Shuwa Arabs, who were extensively employed by the Bornu commanders as auxiliaries and mercenaries, carried firearms, and although these were but miserable little French flintlocks, supplied with bad powder, they caused terror in an enemy, including the Fulani of Mandara.[46] At the battle of Angala,[47] firearms played a larger role than was usual in the West and Central Sudan at this time. Sheikh el-Kanemi ('the Shehu') had taken the field to meet an invasion of eastern Bornu by a Bagirmi army, occupying a position some five miles from that of the enemy. After declining several times to be brought to combat since he could not 'get (the enemy troops) into the situation he wished for', he accepted battle on 28 March 1824 in the plain to the south-east of the town of Angala. At this time he had only a few Arab (Shuwa) musketeers with him, but these he reinforced with some forty Musgo slaves trained to the use of flintlocks. He then placed his musketeers on the flanks of the army and stationed in the centre the two brass 4-pounder guns for which Denham's carpenter, Hillman, had built carriages. He took up his own position behind the guns, surrounded by his Kanembu spearmen, and raising the Prophet's standard awaited the Bagirmi charge. The Bagirmi army, some 5,000 in number with 200 (presumably mounted) chiefs at their head, then advanced 'with great coolness in a solid mass', directing their attack towards the Bornu centre. The 4-pounders opened fire, and with such success that the Bagirmi changed direction and fell upon one of the flanks (Denham does not specify which) of the Bornu. A mêlée ensued whose confusion is fully reflected in Denham's eyewitness account and in which the Bornu and the Bagirmi both suffered heavy casualties. At length the Bagirmi began to give way; soon they were in flight, pursued by the Bornu cavalry. Among the booty which they left behind for the Bornu were 480 horses and nearly 200 of their women. Then, when victory was assured, the Bornu soldiers began to extol the power of their guns (meaning presumably both the cannon and the muskets) which had taken great toll of the enemy. After listening to this for a time, probably with some irritation, el-Kanemi interrupted to tell his troops: 'I lifted my hands and victory was yours!'

In the forested south of West Africa warfare had to be adapted to the difficulty of movement through the tangled undergrowth and to the greatly restricted visibility. Deployment was hardly possible for more than a group of section or platoon size – to use European equivalents for the smallest fighting units. Thus the prevalent form of warfare was the ambush, carried out by concealed parties of warriors whose attacks could be co-ordinated only by means of sound signals, and whose weapons were predominantly those suitable for hand-to-hand fighting – swords and clubs, rather than bows and spears. Outflanking movements through the bush were also practised, the Ashanti being especially renowned for these tactics. Cavalry was greatly hampered if ever it ventured into this close country, becoming easy prey for an enemy experienced in such conditions, as the Fulani of Ilorin found to their cost on their expedition into eastern Yorubaland during the Pole war against the Ijesha (*c.* 1830).[48] Moreover, horses soon fell sick in the tsetse-ridden forests. The introduction of firearms made little difference to the tactics of forest warfare, as is illustrated by the Ashanti defeat of the British-led force during the operations of 1823–4 and by the Ijebu resistance to the British force from Lagos in 1892. Finally, in the small-scale warfare practised by such peoples as the Ibo, it is hardly possible to speak of tactics. Uka writes of the Abam, for example, that their 'Battles were fought in a haphazard manner ... Each fighter followed his own devices ... his main concern being the securing of a human head. Once he attained the coveted prize, he started heading for the war camp.'[48A]

It is not clear what prompted a decision in West African warfare as to whether the main resistance to an invader should take place on the enemy's line of march (this in the forest usually taking the form of an ambush) rather than along the fixed defences of the main towns or camps. Factors such as the nature of the country, the strength of the defences, and the weapons to be used were presumably all considered, and there must also have been a reluctance to yield farmland to enemy occupation and depredation. For example, the professional army of the Yoni, equipped with many guns, engaged the British expedition of 1887 as it moved through the bush, but when the invaders arrived before the stronghold of Robari all resistance collapsed. This, Ijagbemi suggests, was because the stockade around the town could give no protection against the shells, rockets and machine gun fire of the enemy.[48B] It remains puzzling that the Ijebu, who must have been aware of the successful repulse given by their neighbours, the Egba, to the determined attacks of the Dahomeans on the earthen walls of Abeokuta in 1851 and 1864, preferred to throw their whole weight against the British expedition of 1892 into ambushes on the enemy's line of march and an attempt to deny the crossing of the minor river Yemoji, rather than to defend their capital of Ijebu Ode with its walls and ditch.[48C]

It is difficult to find any account of a battle in the forest between West African peoples without any intervention from outside which is sufficiently detailed to throw light on methods of warfare. Perhaps the most valuable is Captain Jones's description of Egba warfare, cited on several previous occasions in this book. Oral tradition about war is especially prone to exaggeration and generally leaves unanswered most of the questions which the military historian would ask. But an account gathered in 1819 by Consul Dupuis of an engagement which had taken place only some twelve years before in the forested hinterland of the Gold Coast does give a vivid if fragmentary picture of operations in this terrain. The protagonists were an Ashanti army and an alliance of Fante warriors, each side being armed with muskets.[49]

During his first day's journey up from the coast on his way to Kumasi, the Ashanti capital, Dupuis came across a pathetic settlement in the forest, 'a few hovels squeezed within a mass of crumbling walls'. He was told that this was all that remained of the once large and prosperous town of Emperou, destroyed as a result of the 'unfortunate neutrality' which it had attempted to observe during the first Ashanti invasion of the Fante country in 1807. When a part of the Ashanti army approached the town, the inhabitants, according to Dupuis's informant, had accepted an offer of military aid from some of their neighbouring Fante kinsmen. So large a force then began to gather in the town that the leaders were encouraged to take the offensive and to interpose their troops between that part of the Ashanti army which was in their vicinity and the Asantehene's head-quarters. Leaving a garrison in Emperou itself, a large force set out to fulfil this plan, 'made a circuitous march to the westward, and fell unexpectedly upon the flank and rear of their adversaries'. The Ashanti quickly rallied, and the attackers withdrew behind the walls of Emperou. But their numbers were so much greater than those of the Ashanti that they decided once again to leave the town and re-engage the enemy. At first the battle seems to have been fairly even, but the discipline of the Ashanti and the 'murderous firing' of their muskets, sustained with 'superior celerity', eventually began to tell against the Fante. At this juncture, as the Fante were still struggling to hold their ground, volleys of musketry assailed their flank and rear, announcing the arrival of the main body of the Ashanti army. Attacked from all sides and cut off from their town, which was already being entered by parties of the enemy, the Fante warriors took to flight. Only some 100 made good their escape, Dupuis was told, the rest being slaughtered, while the town was put to fire.

It is interesting that in this engagement at Emperou the Ashanti showed a superiority in the handling of guns which won them their victory over the more numerous Fante who, as a coastal people, had had a longer acquaintance with firearms. Even as late as this, the Ashanti armies were apparently only partially equipped with muskets, bows being still the major armament of many of the troops. Indeed, the object of their invasion, as has been seen earlier in this chapter, was primarily to gain control of a supply of firearms and a principal result of this campaign was to re-open a trade route between Kumasi and the coastal towns of Cape Coast and Anomabo.[50] Other elements in this Ashanti success which call for notice are the speedy gathering and transmission of local intelligence and the co-ordination of command which, even in this close country, enabled the headquarters to bring quick and effective aid to an independently operating formation of the army when it was threatened by the enemy.

Another episode in this Ashanti invasion of the Fante country in 1806–7 serves to illustrate the part played in war by chance, for after successfully reaching the coast east of the Volta, the army was struck down by smallpox and dysentery.[51] This unforeseen (though not perhaps unforeseeable) occurrence is a reminder that, as Clausewitz writes, 'the mathematical ... nowhere finds any sure basis in the calculations of the Art of War; and that from the outset there is a play of possibilities, probabilities, good and bad luck, which ... makes War of all branches of human activity the most like a gambling game.'[52] Experienced commanders are always well aware of this element of chance, or of the intervention of God or the gods. They make what allowance they can for it in their plans and attempt to placate their deities. Just as mass was celebrated for the armies of Catholic Europe before battles, so, as has been noted earlier, sacrifices were offered by the warriors of West Africa before they took the field. Equally (but not alternatively) the same episode may be taken as illustrating the penalties of failing to observe good hygiene whenever a large body of men is gathered together – a precaution rarely taken by any European armies before the nineteenth century, and not always taken in the present one,[53] and apparently absent from the arrangements of West African armies in precolonial times.

The third form of warfare in West Africa which must be considered is that which involved the use of boats.[54] As described in chapters IV and V above, various types of craft were used on the many waterways of the region, predominant among these being the dug-out canoe. For the crossing of lagoons and rivers, boats were an indispensable auxiliary to the land armies, a feature which is illustrated by the large ferries which Mai Idris Aloma of Bornu built for his troops. They also served as transports, carrying the warriors swiftly and silently towards their objectives and ensuring their subsequent provisioning. The expansion of Songhay under Sonni Ali in the fifteenth century was largely dependent upon such transport down the waters of the Niger (whereas Sonni Ali's successor, Askia Muhammad, undertook long expeditions into the desert, probably using horses). There is much less evidence as to sea-going craft, but at least on the Gold Coast and the Bissagos islands the inhabitants, who were accustomed to fishing several miles out to sea, used their canoes to make seaborne attacks on their neighbours. Apart from this, it is clear that among the peoples bordering the great lagoons and the rivers of the Niger Delta, war canoes were in themselves units of battle from which men attacked each other and whose manoeuvres can be termed a form of naval warfare.

As already mentioned (p. 72 above), the earliest reference to the use of boats in war in this region concerns the operations of a ruler of Kanem on Lake Chad in the thirteenth century. Three hundred years later Mai Idris Aloma of Bornu was said to have attacked the men of Lake Chad by 'enveloping them with many boats', which seems to imply that a battle took place between boats on the water.[55] But for more explicit accounts of naval warfare it is necessary to turn to the history of the coastal and lagoon peoples, especially the Aja, Egun and Awori (a southern Yoruba people) in the eighteenth and nineteenth centuries.

Before the introduction of firearms, the major weapon of these West African navies was probably the throwing spear, with swords and clubs for the close hand-to-hand fighting which was possible on the shallow waters of the lagoons. These weapons continued to be used, for example on the lagoons west of Lagos, as supplementary to firearms down to the mid-nineteenth century. No indications have been found of the use of bows and arrows in the naval warfare of West Africa, but it can be assumed that this popular weapon played its part in a form of warfare to which it was well suited. From the mid-eighteenth century onwards, there are many references to war canoes carrying small cannon – usually six- or four-pounders – in the bows. These were either lashed to the sides of the canoe, and so could only be aimed by pointing the whole canoe at the target, or mounted on a pivot or swivel. In the case of the Owerri canoes seen by Landolphe in the Benin river (p. 113 above), a number of cannon were mounted laterally along the thwarts in addition to the blunderbusses at the prow which, Landolphe claims, could be fired simultaneously.[56] Individual warriors in the war canoes also came to be armed with muskets during the nineteenth century and perhaps earlier. An account of an attack by Lagos canoes on Badagry in 1851 describes the warriors in the boats loading their swivel guns and muskets, pulling inshore, firing, and then returning to their original positions to re-load, a manoeuvre reminiscent of the tactics of musketeers in the land operations of West Africa.

As in land warfare, the ambush was a favourite tactic in the lagoon and river warfare of West Africa. The long grasses and bush of the shore afforded cover where the war canoes lay hidden until they could surprise an enemy. Another form of war was the blockade, of which an example was cited in the last chapter: the stationing of Lagos canoes in 1784 to prevent supplies reaching Badagry in the course of one of the numerous wars between those two states. Descriptions of naval battles in West Africa are rare, but some features may be illustrated from the history of the lagoon peoples and their neighbours. During their invasion of Whydah – described by Dalzel as one of 'the two greatest maritime powers in the neighbourhood of Dahomey', the other being Ardra – in 1726–7, the Dahomeans were much hampered by their lack of canoes.[57] In 1753, however, when they were again fighting the 'old Whydah' (now exiled to the nearby Popo islands but still disputing possession of Whydah beach), the Dahomeans (in Norris's account) 'were provided with some canoes, which enabled them to penetrate into their [the enemy's] country; but it was a mode of warfare which they did not well understand'. They were drawn to a spit of land between the sea and the lagoon, where their retreat was cut off; the Popo (or Egun) allies of the Whydah then opened fire on them 'at their leisure from their canoes'. Norris does not indicate whether this firing was from muskets or cannon (or both), but in any case it was ineffective since the greater part of the enemy 'perished by disease and famine' rather than from wounds.[58]

The war of 1778–82 between Ardra, allied with Dahomey, and Ekpe, provides an even more useful example of canoe warfare, since it includes a battle fought on the water between opposing canoes. In this, writes Dalzel, Ardra, having taken the Ekpe capital, 'repaired in full force to the river, in order to intercept such of the enemy as might attempt to make their escape by water'. The Ekpe, embarking in their canoes, got away to Weme; when the safety of their royal family was assured there, they 're-embarked and went in quest of the Ardra, whom they fought, and put to flight', the Ardra being saved from complete defeat only by the firepower of the two canoes belonging to Antonio Vaz Coelho, a negro trader from Brazil.[59] Another example of a battle on water was the attack on Badagry in March 1845 by the Iso, in the service of the king of Porto Novo, described by a missionary observer. When the Iso canoes were seen approaching Badagry, the Badagrians opened fire on them from the shore with 'a couple of large guns, that was (*sic*) used as signal guns to vessels'. Then 'a small number of Badagry canoes' was launched, and 'between them and the Iso canoes commenced a good deal of firing' – apparently at fairly long range – 'which continued as long as light would enable them to see each other.'[60]

In their practice of war as in their diplomacy, the West Africans were subject to the influence of other cultures, specifically those of the Islamic world and of Western Europe. Something has already been said of those Islamic military influences which began with the arrival of the early travellers across the Sahara, led to the introduction of fire-arms to the Central and West Sudan in the sixteenth century, and culminated in the jihads of the seventeenth to nine-teenth centuries and the somewhat academic treatises on warfare which accompanied them. The European partition of Africa in the last two decades of the nineteenth century had, somewhat like the jihads, been preceded by a long period during which the interventionists had been in-volved in the warfare of West Africa both as active partici-pants and as advisers. The earliest European visitors to the coast, the Portuguese, had founded a series of trade posts stretching from the Saharan island of Arguin to Whydah at the western edge of what was to become the Slave Coast. They were soon joined, and then to a great extent super-seded, by other Europeans – Spaniards, Swedes, Branden-burgers, Courlanders, French, and, most important of all, Dutch and English. At one time or another these were all drawn into the politics and wars of the Africans with whom they traded and alongside whom many of them lived.

The principal European trading posts, eventually some fifty in number, were fortified enclaves, appropriately desig-nated 'castles', of which the prototype, as well as the largest, was Elmina, St George of the Mine, on the Gold Coast.[61] The purpose of these fortifications was to provide shelter for the traders and in troubled time also for the local African community which grew up in their shadow and to repel both European trading rivals and hostile African armies. Despite the strength and skilful disposition of the castles, and the facility with which they could in time of need be provisioned and reinforced from the sea, they were on numerous occasions not only besieged but also assaulted, sometimes successfully, by African troops. Bosman, for example, describes the five years' siege, from 1696 to 1701, of the English fort of Dixcove ('Dikjeschoftt') by the Ahanta, after which an alliance was made between the two sides to cheat all the ships trading there 'by putting sophisticated Gold upon "em"' (but Bosman is a prejudiced witness and the gold there was notoriously impure). The English fort at Sekondi was less fortunate, being overrun about 1698 by a local African force allied with the Dutch from the adjacent fort, and its occupants massacred. Bosman also mentions a furious African attack in 1695 on the 'indifferent large Fort' at 'Little Commany' (Komenda) which he commanded. Treachery within the fort led to the cannon there being put out of action and a dangerous situation was relieved only by the opportune arrival off shore of a Dutch vessel.[62] Another strong castle, Christiansborg near Accra, which was the headquarters for the Danish traders, was captured by a ruse in 1693 and held for some time by a powerful trade chief from the southern Akan state of Akwamu. Thomas Phillips, an English sea-captain, arriving there the following year, reports that he 'bought a five-hand canoe ... of the black general who had surpriz'd and seiz'd the Danes fort here, forced the Danes general to fly to the Dutch to save his life, murder'd his second and several of the soldiers, and now trades with the Dutch interlopers...'[63]

Sometimes deliberately but probably more often against their will, the Europeans on the coast took a hand in the politics and also the wars of the peoples of the coast and their neighbours in the interior. The Dutch were especially prone to involvements of this kind on the Gold Coast. In the early seventeenth century they signed a treaty with the Fante providing that the latter should help them against the Portuguese with the necessary arms by land and with their canoes at sea. They fought wars with Axim in 1649, with Adom in 1659, with Komenda in 1694, and in alliance with their English trading rivals and colleagues with John Konny of Komenda in 1711; in the last, Konny's 900 well-armed and disciplined warriors easily defeated an ill-found and untrained Afro-European force.[64]

Probably the best known of military confrontations between Europeans and West Africans before the era of partition were those which arose from British opposition to the Ashanti drive to the coast. In 1807 the British fort at Anomabu was besieged and its small garrison all but overcome before a truce was arranged.[65] Then in 1824 an Ashanti army of about 10,000 men met an Anglo-African force, consisting mainly of Fante levies with a contingent of allied Wassa, under the command of Sir Charles Mac-Carthy, the British Governor of Cape Coast Castle, at Asamankow. The Ashanti were held for some hours on the banks of the river Adumansu, but towards evening the ammunition of the British-led force ran out; at the same time the Wassa contingent withdrew from the field, abandoning their allies. Realizing their enemy's plight, the Ashanti now forded the river and, despatching a detachment to cut off their retreat, succeeded in overwhelming them. MacCarthy himself, with others of his European colleagues, was wounded, captured, and executed.[66] Two years later, the British took their revenge when a much larger Ashanti army, said to comprise 40,000 warriors, was defeated at Dodowa by contingents from a number of neighbouring states who, with a militia raised by the principal European merchants, formed an army at Accra under the new British Governor at Cape Coast. After a fierce battle lasting some nine hours the Ashanti fled, the Asantehene himself only narrowly escaping capture.[67]

Apart from such cases of direct intervention, the Europeans from time to time gave their advice on the techniques of warfare to the Africans. King Agaja of Dahomey, for example, profited from the presence of the French in their fort at Whydah by learning their views on the art of fort-building,[68] while his English prisoner, Bullfinch Lambe, taught him how to fire his 'great Gun'.[69] The similar services of Denham and Clapperton in Bornu and the Sokoto Caliphate have already been remarked, while British naval officers and missionaries were ready to help the Abeokutans in resisting the Dahomean threat – and the American Baptist, Bowen, took a hand himself in the defence of the town in 1851. In 1892 the French lieutenant Mizon assisted the Emir of Yola in his attack on the pagan town of Kona by lending him two field pieces and the services of his Senegalese sharpshooters.[70] Doubtless many other examples could be added to these.

Towards the end of the nineteenth century relations between the West Africans and the Europeans began to change and to deteriorate. The military and diplomatic problem for West African rulers was, almost suddenly, transformed from that of dealing with white neighbours and visitors whose numbers were small and whose resources were similar to their own, to that of meeting an aggressor whose technical capacity had dramatically increased and whose military effort was now on a much greater scale than any ever before encountered. But the extent of this new threat was not appreciated until too late. The old inter-African animosities persisted, so that there was little combination against the enemy from another continent; indeed, the colonial powers often relied heavily on local African allies and mercenaries in their carving-up of West African territory. Michael Crowder sums up the issue in his Introduction to a symposium on 'West African Resistance':

> West Africa was an agricultural society with limited resources to finance a long-term war, whereas the Europeans came from an industrial society which had, by comparison, infinite resources, in particular the resources of firepower. ... It was Africa's misfortune that the Scramble for Africa occurred at a time of peace in Europe, when Europeans, instead of using their newly acquired weapons on each other, used them on Africa.[71]

VIII. A summing-up

The object of the preceding chapters has been to examine the war and diplomacy of pre-colonial West Africa as different phases in the continuing function of international relations, rather than as contrasting conditions. Every state and every society, however self-absorbed, must conduct, however reluctantly or passively, such relations. Conflict and harmony of interests, incompatibility and compatibility of aims, these separate peoples and bring them together, cause wars and weld alliances, and in sum make up the pattern which gives meaning to political history – which is the history which matters above all. To write a history of relations between any nations and peoples requires an analysis into terms of cause and effect of such conflict and such harmony. The intention of the present book has been different, both more modest and more fundamental, in that it sets out primarily to examine the means by which the two activities of war and of peaceful international relations were carried on in a particular (though perhaps too vaguely specified) period of time and region.

A study of pre-colonial West Africa on these lines must necessarily embrace a great diversity of peoples and of political organizations. This in turn has led, almost against the writer's will, to an attempt which some will think still more (and more unsuitably) ambitious: to draw together the practices in war and diplomacy found over this field, so wide and varied, by postulating and testing generalizations. The method has usually been to put forward a statement about the conduct of diplomatic relations or of war under particular circumstances which evolved first as a hypothesis or one of several hypotheses in the writer's mind, and then to offer examples which seem to substantiate it. To the reader this may have seemed a putting of carts before horses, of theories before evidence, and the writer is himself acutely aware that two or three examples cannot validate a generalization, much less establish a 'law'. It must, therefore, be admitted at once that the generalizations which occur throughout this book amount to no more than possibility or sometimes, at best, probability statements, and that they are submitted simply as suggestions that this was the pattern of war and peace among West Africans in these centuries. All that can be 'proved' from the evidence offered, or, to accept Oakeshott's phrase, all that the evidence obliges us to believe, is the unsurprising conclusion that in large parts of West Africa before the partition of the region among European powers, international relations in peace and war were carried on in a more or less recognizable fashion, and, to go a little further, in a coherent and rational manner which showed itself capable under favourable conditions of leading to political, economic and technical improvements in society.

In attempting to sum up what has gone before, it becomes necessary to re-state and make explicit a few of the more important of the generalizations referred to in the last paragraph. First among these should be, perhaps, the link which has been claimed to exist between the formation and growth of states in West Africa and the processes of war and diplomacy. That most of the examples adduced have been taken from the histories of those societies known as states indicates that it was only such societies which had the resources to practise these activities on a scale which usually left a record behind. But beyond this it seems that war was itself a force, and probably the greatest force, in the creation of statehood, so that the attainment of statehood – that intangible concept – was fostered as well as characterized by the development of means by which the interests of the state could be forwarded. Examples of offensive war as the major founding element in states abound: medieval Mali, expanding along the Niger under Sonni Ali and across the desert under Askia Muhammad; Dahomey's rise to power and conquest of her coastal neighbours under Agaja; the creation of the Sokoto Caliphate by a revolution from within: such examples come readily to mind. Perhaps of still greater interest in this connexion is the suggestion, convincingly argued by Professor Abdullahi Smith, that the Hausa states were built up from the strong defensive core of the walled *birni*, as the response to the aggressions of others. In these examples, the role of diplomacy is less obvious than that of war, though these were all states which attributed importance to the role of their peaceful agents abroad. One state, Ashanti, seems to have been outstanding in the use which it made of diplomacy as an alternative to war in the prosecution of its claims, but there are many other cases where diplomatic activities have brought peoples together in alliances and unions for their greater protection and for the fulfilment of their common aims.

In the maintenance of states, as in their formation, war and diplomacy retained their importance and continued to be agents of change. The need to raise an army of a size beyond the capacity of a single man to control led to the creation of war chieftaincies; the improvement of weapons and the introduction of new ones required skilled and experienced men to handle them, and thus a class of professional, or full-time, and semi-professional warriors arose. Such distinctions proliferated and became ever more complex. The chiefs soon extended their role into civil government (as did, for example, the Kontihene and Adontehene of Ashanti), while society adjusted itself to the changes brought about by the changing techniques and demands of warfare, changes which have been described here in accelerated and simplified form but which in West Africa took place usually hesitantly and at a very slow pace. War not only created chiefs; it also strengthened monarchies as the crises undergone by the state called for the greater coercion of citizens and an intensification of the nation's military and economic efforts. Here Dahomey, as so often, provides an apt example. The rise of the Dahomean state and monarchy (the two being inseparable) illustrates the link identified by Andrzejewski between the waging of aggressive war and the growth of centralized rule – though a contrary example may be adduced from the Sokoto Caliphate. On the other hand, the collapse of Oyo and the immigration of many of its peoples into newly independent towns in the security of the southern forests illustrate the complementary relation between defensive war and the dispersion of political power. Meanwhile, diplomacy was valued as a less costly means of forwarding national interest. Like war, it gave rise to new classes of leaders and state servants – men who could combine or alternate insinuating manners with airs of confidence and authority, who could orate and debate and persuade, had their own kind of tactical skill, and, as the requirement arose, could master foreign tongues and the arts of reading and writing.

On the technical side, much has been claimed for the influence of war in forwarding 'progress'. The inventions, new techniques, and forms of transport fostered by war were rarely of importance only for destruction. The throwing stick, the javelin, the bow and arrow, and the handgun were all implements useful for hunting and which added to the food resources of a people adept at their use. Horses and other animals were at least as much used for civilian communication as for warfare, as were boats – the last being also a valuable adjunct to fishing. Other writers, mostly those studying African societies in general rather than their individual histories, have advanced more detailed generalizations, some of which cannot be accepted until more evidence is forthcoming and some of which seem to be contradicted by such evidence as is available. Goody, for example, claims that 'the sword was never of great importance (except ritually)' in Africa (south of the Sahara is implied), which is certainly untrue of several of the forest armies of West Africa; that in West Africa 'horses were the possession of a politically dominant estate that was usually of immigrant origin', the second part of which statement is apparently substantiated in some cases, such as that of the Dagomba, and belied in others, for example, that of the Oyo; that 'the horse was the basis of military organization' in the savannah, the gun in the forest states, a neat contrast but one which cannot be reconciled with the place of the cavalry in seventeenth- and eighteenth-century Oyo or of the archers and spearmen and later the musketeers in many of the armies of the savannah.[1] Another social anthropologist, Lloyd, writes that 'With tribal methods of warfare', as practised by the Yoruba in the nineteenth century, 'the defence of a community is safeguarded more by its size than by any other factor',[2] a statement which amounts to no more than the claim that all things being equal (which they never are), a more numerous force will defeat a less numerous one – which is a preposterous banality.

War is sometimes held to be inimical to the development and even the maintenance of a nation's economy. That this is by no means always so is shown by the application in civil life of many of the technological improvements made by war, in the discovery of substitutes for goods no longer available, and in the opening-up of new markets and sources of production and supply. Among the West African evidence, the examples of Ibadan illustrates the tremendous impetus given to the economy of this Yoruba town by militarism. Yet, as Mrs Awe has recently shown, militarism also imposed limits beyond which economic growth of this kind could not go.[3]

On the diplomatic side, the changes wrought by contact among different peoples in West Africa were less tangible and much less obvious than those wrought by war, but it seems permissible to remark again on the extension of communication by the learning of foreign tongues and customs and by the evolution of the formulas and protocol of diplomacy. Thus diplomacy increased the knowledge of foreign cultures and techniques and led to the adoption of such of these as were found desirable and useful. The practice of diplomacy itself presumably spread in this way, and certainly trade was encouraged and facilitated.

Finally, it has been suggested that diplomacy and, paradoxically, war too led to the spread of laws and customs on a scale which is held to have created a West African 'international law' – which is to read backwards Hoebel's dictum that 'international law, so-called, is but primitive law on the world level'.[4] Many similarities in African customary law which have been observed over a wide area can be assumed to derive from a similarity in physical and cultural conditions rather than from deliberate modifications or borrowings from the practice of neighbours and others at a great distance. Yet the customs and forms of international usage continued to evolve to the end of this pre-colonial period and in this evolution much must have been due to interchange of ideas among different peoples. Women and children were often spared from massacre even if their fate was then usually enslavement, declarations of war were perhaps more common than the surprise attack, and prisoners could in some cases be redeemed for payment, while on the diplomatic side the persons and property of the diplomatists were usually considered sacrosanct and the treaties which they concluded were 'meant to be kept'. Objective evidence is lacking that such ameliorations in the ways of men were a continuous process or were more prevalent at the end of the period in West Africa under consideration than at the beginning (though research directed at particular topic may someday yield more information), and it is very far from the intention here to suggest that any law of 'progress' was in operation. It is clear that the civilizing influences of both military organization and diplomacy were considerable and had an impetus of their own. It is equally clear that the gains of any generation, however painfully achieved, could be and often were discarded by their successors. (It was, after all, the professionalism of soldiers which outlawed that affront to humanity, the 'guerilla', the evil of whose activities is nevertheless less widely recognized now than a few years ago.)

Thus, on the eve of the colonial partition, much of West Africa, and especially that greater part which was organized into separate, distinctive states, existed within a system of international relations which, however embryonic when measured against the by no means infallible or final standards of Europe or Asia, was adapted to its own needs, flexible, and time-honoured. Neither on the coast nor in the savannah nor in the deepest forest did these people live in isolation from each other. Their peaceable relations were conducted in obedience to a customary law which was widely accepted, comparable in many ways to that which obtained in Western Europe and elsewhere, and capable of assimilating without being essentially changed by external influences. They were also the prey of frequent wars. *Pace* Captain Jones at Ijaye, the factual content of whose report contradicts his own prejudices, these were not merely 'the irregular marching and skirmishing of a barbarous horde'.[5] They were conducted by national armies whose organization, not to be easily understood by an outsider, was of considerable complexity and, though often too rigidly maintained, was by no means incapable of adaptation to new conditions or of absorption of new methods of warfare.

Modernization, though not necessarily 'westernization' as some writers have called the process, made rapid advances during the last hundred years of the pre-colonial age in warfare and, to a lesser extent, in the methods of peaceful negotiation. Yet this was confined within narrow limits. Many West Africans continued until the end of the nineteenth century to fight their wars with the weapons which preceded firearms, while the adoption of breech-loading rifles by the armies of the advanced states was unaccompanied by equally striking changes in strategy, tactics or logistics. Similarly, increasing contact with the Islamic world and with Western Europe led to the Development of a more refined and alert diplomacy, and yet again this affected only a minority of the peoples and was as ineffective as military means in resisting the advance of the European powers who partitioned Africa at the end of this period.

Notes

(For all bibliographical details, see Bibliography)

Chapter I

1. Fortes and Evans-Pritchard (1940), 5–23; Paula Brown (1951), *passim*; Lloyd (1971), 1–8; Law (1973), *passim*. .
1A. J. C. Yoder, 'Fly and Elephant parties: political polarization in Dahomey, 1840–1870', *J. African History*, XV, 3 (1974), has recently argued that the absolute power of the Dahomean monarchy has been exaggerated. He points to the development in the mid-nineteenth century of two groups advocating different policies for the kingdom at the Annual Customs.
2. Nicolson, 52.
3. In Gilbert (1966), 23–33.
4. Wright, 1284.
5. For example, Farrer, 359–60.
6. Wright, 98–9.
7. Elias (1956), v.
8. Allott, 64–71.
9. In Larsen, Jenks and others (1965), 210–22.
10. ibid., 220.
11. See, for example, Rodney (1970), 87–8. Sundström, 5–11, illustrates broadly similar practices in taxation, including customs and tolls, in Guinea.
12. Diamond (1971), 4, 41.
13. ibid., chapters 7, 25, and Diamond (1965), *passim*.
14. See Horton in Ajayi and Crowder, I, chapter 3, for a convenient summary of what constitutes a 'stateless' society and of the role of 'secret' societies.
15. Numelin, (1950), 72.
16. See pp. 53–4 below.
17. For example, the Bini conquered Lagos and established a Bini dynasty there in (probably) the sixteenth century without disturbing Awori landownership. Another example is the Oyo conquest of the Egba whose rulers were apparently allowed to retain their thrones.

18. Mabogunje in Ajayi and Crowder, I, 6; H. J. Fisher (1972), 368–72.
19. Turney-High, 19, claims that the bow was ineffective in the long grass areas of Africa and was more used in the forests. Evidence from West Africa, for which see chapter IV below, does not support this.
20. Hunwick in Ajayi and Crowder, I, 212–13.

Chapter II

1. Trotsky, *Mein Leben* (Berlin, 1930), 327, cited by G. A. Craig and F. Gilbert, *The Diplomats, 1919–1939* (Princeton, 1953; 1965 ed.), I, 235. Similarly, Trotsky began his task at the Narkomindel by saying: 'I will issue some revolutionary proclamations to the people and then shut up shop.'
2. Numelin, 13, citing G. C. Wheeler, *The Tribe, and Intertribal Relations in Australia* (London, 1910).
3. Hodgkin in Ingham. Al-Bakri's famous description of Ghana is cited by Levtzion in Ajayi and Crowder, I, 126–7; Levtzion (1973), 41; Leo Africanus, 825 in Book VII.
4. Ryder (1969), 73. Two of the missionaries were Franciscans and the third a member of the Order of Christ. The letter is translated in full in Ryder (1961).
5. Ryder (1969, 113.
6. Snelgrave, 5–6.
7. ibid., 135. See also Akinjogbin (1968), 91–2.
8. Daaku, 124.
9. Tenkorang, 5.
10. Fynn, 84–6.
11. ibid., 124.
12. Akintoye, 146.
13. See Goody, 39–42.
14. See Lovejoy, *passim*.
15. ibid., 539.
16. Mattingly, 63–4, cautiously admits the place of commercial agents in the history of diplomacy. He concludes: 'For Venice, anyway, a case might be made for her consuls having been the precursors of her resident ambassadors.'
17. ibid., 10, 66–76, 146–7. Mattingly also postulates (57–8) a connexion between the development of mercenary armies, and consequently of wars more difficult for statesmen to control, and the development of permanent diplomacy.

18. Numelin, 128, claims – wrongly, if these examples are accepted – that permanent envoys are found only among the Indians of North America and 'historical' peoples.
19. Leo Africanus, 830. See also Hunwick in Ajayi and Crowder, I, 214–15. There is some doubt, however, as to whether Kano was in fact ever subject to Songhay.
20. Daaku, 69, 159.
21. Dupuis, xi, xiii, xviii; Arhin, *passim*; Fynn, 121, 142. Arhin maintains (76–7), unconvincingly, that the Ashanti presence in such places as Dagomba was 'not political'.
22. Levtzion (1965), 103–8.
23. Argyle, 25, citing Dalzel, 175, 205, 213. These references suggest but do not confirm that there was a permanent Oyo mission in Dahomey.
24. Smith (1969), 50, 162–3; Ryder (1969), 21; Akintoye, 98.
25. Abdullahi Smith in Ajayi and Crowder, I, 188. Originally the *sarkin turawa* was probably himself an Arab or Berber.
26. A. H. M. Kirk-Greene in a private communication.
27. M. G. Smith, 43, 55; Barth, III, 260, 405, 424. The first Mestrema was a war prisoner of the Sultan of Bornu.
28. Davies, 5. The first Kunkuma was a hostage taken by the Ashanti after the treaty of Jarapanga.
29. Bowen, 99; Newbury, 47. See also Fynn, 121, for the three Ashanti officials posted to the Danish, Dutch and English quarters of Accra respectively in 1776.
30. Burton (1864; 1966 ed.), 88, 124, 149, 189. When received by king Glele, Burton was sponsored by the Meu and his 'dependent, the English landlord'.
31. Dorjahn and Fyfe, 391–7.
32. Turney-High, 246, claims that in Africa 'Foreign relations were ... always the royal prerogative wherever monarchy existed', but gives no evidence.
33. Kyerematen, 28.
34. Johnson, 513.
35. Caseley Hayford, 37. The proverb in Fante runs: *wori bobo ingwa*.
36. Burton (1864; 1966 ed.), 344.
37. Bowdich, 252; Wilks in Forde and Kaberry, 212–13. Hagan has traced the beginnings of the administrative changes beyond the reign of Osei Kwadwo, 1764–1777, to that of Opoku Ware, *c*. 1720–*c*. 1750.
38. Bontinck, 48–9.

39. Bosman, 75.
40. Meek (1937), 244.
41. Dupuis, 168.
42. Bowdich, 248–9, 257; Kyerematen, 92. Agyei is the subject of Story V in *Stories of Strange Lands* (London, 1835), 152–79, by Mrs R. Lee, formerly Mrs Bowdich.
43. Astley, II, 68, 73. Sometimes Europeans were misled as to the status of their African visitors, as may have happened in the case of 'prince' Aniaba, received by Louis XIV in the character of heir to the throne of Assinie and whose rank was later questioned: see P. Roussier, *L'Etablissement d'Issiny 1687–1702* (Paris, 1935).
44. Nys, 12, points out that a Venetian ambassador's expenses were often three or four times the sum allowed by the Signory.
45. Johnson, 60–3, 67, 468. See also Law (1971), 75–80.
46. Johnson, 211, 468–70, 531, 627. The first describes a peace conference at Ikoyi between the Alafin and his rebellious chiefs during the last years of the Old Oyo kingdom. The Ilari who represented the Alafin was Kafilegboin. This was taken by the chiefs to imply that the Alafin had no genuine wish to reach agreement, and they broke off negotiations. The other references concern the objection by Governor Carter of Lagos in 1893 to dealing with Oba ko she tan, who had taken part in abortive negotiations to end the Yoruba wars in 1881. The Alafin sent Oba l'olu in his place. Johnson, 72, apparently errs in describing the Oyo chief known as Laguna as 'the state ambassador in critical times'.
47. See Fraser and Cole, 220.
48. Dalzel, 123, 154, 171; Argyle, 68–9. Akinjogbin, 118, writes that king Tegbesu (1740–1774) introduced Oyo court institutions, including officials corresponding to the Ilari, into Dahomey.
49. Burton, 87; Adandé, *passim*; Kyerematen, 92–6 and illustrations 94, 95, 97. For similar devices – the chalking of faces, the wearing of palm leaves, bells, or white baft – among the smaller-scale societies of south-east Nigeria, see Talbot (1923), 239; Talbot (1926), 825, 837; Fraser and Cole, 310.
50. Bosman, 194; Bowdich, 294; McCulloch, 29–33; Willett (1971), 184–5; Fraser and Cole, 306. For the Dan view of the political role of masks, see Horton in Ajayi and Crowder, I, 103, fn. 36.

51. Mattingly, 42–4, writes that although various forms of immunity had been established in Europe by the fifteenth century for the ambassador and his suite, there were so many exceptions that an international law of diplomacy cannot be said to have existed before the middle of the seventeenth century. Nys, 33–4, considers that the law became so exaggerated in the seventeenth century as to diminish the principle of national sovereignty, a development for which he blames Grotius.

52. Ajusafe, 21.

53. For examples of receptions, see Leo Africanus, 824; Barbot, 128, 370, 396; Landolphe, 104–8; Denham and Clapperton, I, 64–6, 67–8, 78–9; Burton, 77, 143–50. For examples of safe-conducts, see Sundström, 15–16.

54. Dapper, 266 (with engraving).

55. Bosman, 126–7.

56. Barbot, 56. The same king, when receiving the French factor at Gorée, removed the hat worn by an accompanying friar, refused to return it, and eventually sent a young slave in return.

57. Claridge, I, 186.

58. Tremearne, 175.

59. This was precipitated, according to Mattingly, 266–7, by the embassy chapel question in the sixteenth century, after the Reformation, and was explicitly stated by Grotius in *De Jure ac Pacis*, II, xviii, para. 4, no. 8. The principle was challenged by a London divorce court in May 1972 (*The Daily Telegraph*, 12 May 1972).

60. For a note on Yoruba sanctuaries, see R. Smith in Crowder (1971), 198. Davie, 208–9, and Numelin, 228, 278, draw attention to the religious basis of sanctuary. Markets were often held under similar protection. Diamond (1971), 343, mentions that 'villages of refuge' were provided under Anglo-Saxon law where there existed an important shrine or temple.

61. Dalzel, viii; Argyle, 64, 72. Snelgrave, 39, writes that courtiers whispered their messages for the king 'into the Ear of an old Woman, who went to the King; and having received his answer, she returned with it to them.' This was probably one of the female officials known as the 'mothers' of the higher male officials and who, according to Herskovits, cited by Argyle, formed a part of the royal system of control.

62. Dapper, 267.

63. Johnson, 592–3.
64. Labat (1730), II, 335–6.
65. Ibn Fartua, 21–2.
66. Bosman, 396–8; Dupuis, 21, 254, 256; Robertson, 148; Fynn, 111; Boahen, 3.
67. Akintoye, 89, 94. It is also alleged that Fabunmi of Imesi Igbodo habitually executed the messengers sent by the Are Latosisa of Ibadan.
68. Al-Bakri, cited by Levtzion in Ajayi and Crowder, I, 126.
69. Ibn Battuta, 320, 328.
70. Cited in Ryder (1969), 30.
71. Barbot, 128; R. Smith (1969 b), 21.
72. Tully, 206–7.
73. Dapper, 305.
74. For example, see Antera Duke's diary in Forde, also Labat (1730), II, chapter X on the interpreters at Allada. Labat adds the picturesque touch that the slightest mistake cost the interpreters their heads. Labarthe, 70, complains that the English were at a great advantage in trading on the Guinea coast as knowledge of their language was so widespread that they did not need interpreters. At about the same time – the early nineteenth century – Robertson, vi, deplored the lack of knowledge of English by the natives of West Africa. Landolphe, II, 43, apparently had to communicate through English at the court of Benin, which must have galled this Anglophobe.
75. In Yoruba, for example, the word *ajele* for the resident representatives of Ibadan was apparently introduced for this purpose. It was applied in the mid-nineteenth century to the British consuls at Lagos and later to British Residents and District Officers. In Yoruba-language newspapers today *olotu* ('head') is generally used to translate 'consul', while 'ambassador' or 'envoy' is *ashoju*, a title used by chiefs who have the special duty of representing their king. State messengers in earlier times were sometimes known as *iko*.
76. Little, *passim*.
77. Personal communication from Professor J. D. Hargeaves.
78. See Bloxham, 295–9. This and other examples are now in the Pitt-Rivers Museum, Oxford, case 24.
79. Snelgrave, 9.
80. Akinjogbin, citing Le Herissé and Hazoume.
81. Argyle, 83–4.

82. Bowdich, 161, for example, describes the close watch kept on his mission in Kumasi.
83. Bowdich, 294.
84. Astley, II, 66. See also Rodney, 195–6.
85. Dupuis, 166, 213.
86. Bowdich, 123.
87. For examples, see Snelgrave, 58; Dalzel, 157.
88. Wilks in Ajayi and Crowder, I, 352, 385.
89. Barbot, 370. But this custom, even if accurately reported, had certainly died out by the latter part of the eighteenth century.
90. Ibn Khaldun, II, 346–7; Dupuis, 174; Davies, 5. The leopards but not the ambassadors reached England. For other examples, see Sundström, 1–5, and Verger (1970), *passim*.
91. Dalzel, 180–6. Akinjogbin, 167, treats Dalzel's allegation of Dahomean intervention with some scepticism.
92. Dalzel, 193–6.
93. Adeleye in Ajayi and Crowder, I, 510, citing the *Kano Chronicle*; Fisher in Gray, 69.
94. Snelgrave, 135; Dalzel, 74; Akinjogbin, 90–2, 123–4.
95. Fynn, 86, 105; Davies, 5.
96. Adeleye (1971), 69.
97. Bohannan, 30–1.
98. Labat (1730), II, 107–9. The six articles of the alleged treaty are set out on 109–13.
99. In Larsen, 222.
100. Bohannan, 30–1. Barbot, 338, refers to the mixing of blood in the solemnization of contracts. See also Basden, 387–8.
101. Labat (1730), I, 179, 208; Atkins, 73.
102. Snelgrave, 135; Tenkorang, 13. For examples of hostages, see Barbot, 298; Davies, 5; Daaku, 170; Sundström, 20–2. In this connexion, a maxim by Folarin, 54, is significant. Writing of contract (*adehun* in Yoruba), he says: 'There is no agreement without consideration, in money, kind or otherwise.'
103. Smith (1969), 41–2.
104. Bosman, 149.
105. This is to go less far than J. S. Trimingham, 147–8, who writes of Islamic laws as being generally subordinate to African customary law. It is not clear if he intends this to apply even to such Islamized states as the Fulani emirates of Hausaland.

106. In the Sokoto Caliphate all foreigners who submitted to the protection of the Caliph and yet retained their own customs and religion, including Hausa pagans, were considered *dhimmi*; see Last, 67.

107. Trimingham, 139.

108. Ibn Fartua (2), 19.

109. Khadduri, chapter XXI. Unfortunately he does not deal with diplomatic relations between the Muslim states.

110. Ibn Khaldun, II, 346–7.

111. Levtzion in Ajayi and Crowder, I, 134, 155. See also Levtzion (1973), 214–17.

112. Hunwick in Ajayi and Crowder, I, 235–7. See also Hunwick (1971).

113. Hunwick in Ajayi and Crowder, I, 206–12; Adeleye, also in Ajayi and Crowder, I, 498, 502–5; Martin, *passim*. The French surgeon's evidence cited by Hunwick (as also by Barth) is from the *Bulletin de la Société de Géographie de Paris*, 3e série, xi–xii, 1849.

114. Bivar and Hiskett, 105.

115. Leo Africanus, Book VII, 825.

116. Last, *passim*, especially 92, 190–5.

117. Bowdich, 90–1, 232, 296; Dupuis, 112, 229. See also Wilks in Forde and Kaberry, 218.

118. For the identification of some of these Muslims, see Levtzion (1965), 99–119. According to Dupuis, 245, the previous Asantehene, Osei Kwame, had been deposed for planning the extension of Islamic practices in Ashanti.

119. Bowdich, 296.

120. Landolphe, II, 85–6.

121. Burton (1864; 1966 ed.), 157–9. There is a reference to the presence of 'Malays', that is, Muslims, in Dahomey in Norris, 102–3.

122. Ryder (1969), 31–2, 45–6, 48, 72. Ryder also mentions (73, fn. 1) that the ruler of Labedde sent an envoy to Portugal in 1520.

123. Bontinck, *passim*. Nys, 19, confirms that such embassies of obedience to a new Pope were especially ceremonious.

124. Akinjogbin, 27–31. For the embassy to France, see Astley, II, 73–7.

125. Daaku, 12, 67, 80; Fynn, 136.

126. Verger, chapter VII. This also describes tentative attempts by the king of Lagos in 1770 and 1807 to make diplomatic

contact with the Portuguese, and an embassy of 1810 from the king of Ardra (Porto Novo) to the Portuguese at Rio de Janeiro. Johnson, 168, records that Obalokun, an Alafin of Oyo who probably reigned in the early seventeenth century, 'was said to be in friendly relations with the king of France (probably Portugal) with whom he had direct communication. It is said that the King sent 800 messengers with presents to that European sovereign, but that they were never heard of again.' There is no confirmation of the tradition.

127. Mentioned by Bontinck, 37, 38.

128. Rodney, 12, 80.

129. See E. Collins and J. D. Fage (1967), both *passim*, for assessments of the *Mission* and the *Journal* of Bowdich and Dupuis respectively.

130. Dupuis, Part II, xiii. Dupuis also criticizes his interpreter, a Fante who gave offence at Kumasi by his ignorance of 'diplomatic courtesy'.

131. Daaku, 67–8, 119, 165; Fynn, 90, 95, 103, 107, 132, 144, 156–9; Ijagbemi, 243.

132. Daaku, 55, 57, 78, 184.

133. Quoted in Ryder (1969), 140–1.

134. Akinjogbin in Ajayi and Crowder, I, 331.

135. Kirk-Greene (1959) gives the text of Barth's treaty of 1852 between Britain and Bornu and reproduces the final clauses, signatures and seals in his edition of *Barth's Travels in Nigeria* (London, 1962), 149. Boahen in Ingham, 5–6; Adeleye, 121, 131–5.

136. Markov and Seebald, *passim*.

137. Khadduri, 286. The Sultan admitted permanent representatives to his capital in the sixteenth century, but it was not until 1792 that he appointed his own resident ambassadors to Paris, London, Vienna, and Berlin. A. Claridge, I, 494.

138. Labarthe, 165.

Chapter III

1. Fynn, 100.

2. For example, see Ibn Fartua, 18.

3. Dan Fodio, 451; Khadduri, 105. Denham, in Denham and Clapperton (D), 149, writes of Ramadan as being the usual season for military expeditions, clearly a misunderstanding.

See also Ibn Fartua, 96, which describes how Idris Aloma continued operations at Amsaka despite the appearance of the new moon of Ramadan. (In October 1973 the Grand Sheikh of Al-Azhar, referring to the war between Egypt, Syria and Israel, said that Muslims fighting in Ramadan were 'sure of Allah's support' and the souls of those killed would rise directly to heaven. *The Times*, 11 October 1973)

4. Ibn Fartua (2), 15; Norris, 15.
5. Bosman, 183. Labat (1730), I, 368, writes similarly that West African wars rarely consisted of more than a single campaign, the usual duration of which was a mere four days; nevertheless some of the Fante wars had continued for four years.
6. M. G. Smith in Forde and Kaberry, 111.
7. Davie, 55; Quincy Wright, 63–5.
8. Tremearne, 185; Meek (1937), 242; Afigbo, 79.
9. For an examination of the problems of causation in a West African setting, see Robert Smith (1971, a).
10. In Gilbert (ed.), 22–4.
11. Ibn Fartua, 5, 55; Norris, 26; Tenkorang, 1–2 and *passim*; Boahen in Ajayi and Crowder, II, 170.
12. Turney-High, 179.
13. Leo Africanus, 825, 833.
14. Khadduri, 130–1.
15. In Denham and Clapperton (D), 149. See also Barth, III, 118, 233; Brenner, 53.
16. Dapper, 307; Barbot, 39. See also Bosman, 183; Labat (1728), 217; Labat (1730), I, 362: II, 238.
17. In Ajayi and Smith, 124. See also R. Smith (1969, a), 157.
18. Hopkins (1968), 591.
19. Ajayi and Austen, with rejoinder by Hopkins (1972), 303–12.
20. See Crowder's narrative, edited by Ajayi, in Curtin (1967), 289–316.
21. Ajayi and Smith, 51–2; Akintoye, 150.
22. Argyle, 81–3.
23. See Curtin (1971), *passim*, and Levtzion in Gray, 199–216.
24. Echenberg, 247; Hunwick in Ajayi and Crowder, I, 230; Njeuma, 5.
25. Bosman, 151–2; Barbot, 297. King Glele of Dahomey interpreted the earthquake at Accra in July 1862 as a warning from the invisible world that he should make war on the Egba: Ajayi and Smith, 116.
26. Argyle, 83.

27. Meek (1931), 351–2, for the Jukun; Ajayi and Smith, 29, for the Yoruba; Talbot (1923), 233, for the Ibibio; Uka, 79, for the Abam Ibo.
28. Talbot (1923), 234; Meek (1937), 257 fn.
29. Fernandes in Hargreaves (1969), 16; Bowdich, 271–2; Davies, 10; Meek (1931), 305; Goody in Forde and Kaberry, 186–7; Fynn, 137. Charms continued to be worn by soldiers serving in the British-raised West African Frontier Force, as may be seen from a photograph (apparently of about 1900) reproduced in Kirk-Greene (1964), 140.
30. Kyerematen, 12; Ajayi and Smith, plate opposite 2.
31. Fisher (1971), *passim*.
32. Ibn Fartua (2), 70.
33. Snelgrave, 10. See also 143–4 below.
34. Meek (1937), 243.
35. Fisher in Gray, 69; Bosman, 181; Atkins, 200; Barbot, 58–9; W. Smith, 218.
36. Dalzel, 167; Johnson, 604.
37. Dapper, 311; Bosman, 457; Barbot, 370.
38. Bosman, 65.
39. Dan Fodio, 503–5.
40. Bosman, 76; W. Smith, 186; Ajayi and Smith, 50–1; Last, 31.
41. Kyerematen, 27–8.
42. Labat (1730), I, 365; Norris, 37.
43. Tremearne, 183; Talbot (1923), 237; Talbot (1926), III, 822.
44. Meek (1937), 243; Basden, 378.
45. Khadduri, 102–6; dan Fodio, 451 and *passim*.
46. *Tarikh*, 169.
47. Bosman, 110; Davies, 10; Meek (1931), 352; Talbot (1926), III, 823.
48. Dapper, 302; Labat (1730), I, 370–1; Ajayi and Smith, 50–1. Cf. the treatment of sword cuts in medieval England by the application of milfoil.
49. Daaku, 32.
50. Khadduri, 127–31; Fisher in Gray, 78, 87, 102, 122. Fisher mentions the case of a captive Bornu prince who in 1667 was a slave in Tripoli. See M. G. Smith (1960), for the sharing out of prisoners in the Habe (pre-Fulani) kingdom of Zazzau.
51. Ibn Fartua, 28, 43; Barth, III, 194; Snelgrave, 31; Dalzel, 202; Dupuis, 87.
52. Fynn, 172.
53. Ibn Fartua (2), 26; Akinjogbin (1968), 110–11; Davies, 5. One

of these chiefs was the Kunkuma (see 17 above), and the other the Demonkum.

54. Bosman, 183; Labat (1730), II, 239; Ajayi and Smith, 52, 120–1.
55. See M. G. Smith in Forde and Kaberry, 114, for an account of the role of booty in the wars of the Hausa kingdom of Maradi, Fisher in Gray, 70, and Davie, 81, 96, for general comments on booty.
56. Khadduri, chapter XI; dan Fodio, 116, chapters 33–6.
57. Nadel, 112; Last, 106; Adeleye, 86–7; M. G. Smith (1960), 99; Barth, III, 252–3; Denham in Denham and Clapperton (D), 250.
58. Tremearne (1912), 165, 185–6; Meek (1937), 42, 47, 246.

Chapter IV

1. Ibn Fartua, 16; Dapper, 302; Bosman, 70; Barbot, 57, 184; Dalzel, x; W. Smith, 217; Robertson, 153, 271; Meek (1931), 351.
2. See, for example, Ajayi and Smith, 15, 80, 133.
3. Abdullahi Smith in Ajayi and Crowder, I, 147; Fisher in Gray, 99.
4. For example, see Johnson, 135.
5. Ibn Fartua (2), 41; Labat (1730), I, 367; Denham and Clapperton (D), 132; Tremearne, 184; Rodney, 65; Ajayi and Smith, 22, 39, 87, 135. See also Davie, chapter III, and Turney-High, 152.
6. Bosman, 394. The statement is repeated by Barbot, Norris and Labat.
7. At Poitiers in 1356 the French army consisted of about 20,000 men, the Black Prince's of some 7,000. At Bosworth Field, 1485, the Yorkists numbered about 10,000, the Tudor army about 5,000. The Royalist army at Marston Moor numbered 17,000 against 26,000 rebels.
8. Cited by Fage (1957), 80–3.
9. Leo Africanus, 825, 831, 833.
10. *Tarikh*, 219. Rainero, 227–8, 239, attempts to reconcile these figures with those given by other authorities and concludes that the Askia's army at Tondibi probably totalled 30,000 to 40,000 men.
11. Bosman, 181. See also Dapper, 287, 311.
12. Burton (1864; 1966 ed.), 264; Argyle, 89; Akinjogbin, 147.

13. Last, 72; Smaldone (1970), 38; private communication from Mr J. Lavers. See also Barth, II, 144 for the Kano army, and III, 196–7, where he estimates the number of cavalry in a Bornu expedition at 10,000 with a still greater number of infantry.
14. Bowdich, 300, 315–6; Dupuis, xxxviii.
15. Law (1971), 337.
16. Ajayi and Smith, 86–7, 89. Unfortunately, Captain Jones, the British officer sent to observe the war, makes only an ambiguous reference (ibid., 133–4) to the numbers of the Egba army. He implies that each Egba 'head chief' brought into the field the equivalent of a brigade, that is, some 4,000 men.
17. Akintoye, 116. No authority is given for these figures, but they apparently come from missionary sources.
18. R. Smith in Crowder (1971), 179–90; Legassick, 109, 114.
19. Davie, 32, gives examples of the use of female troops in the Canaries, Hawaii, and various parts of South America.
20. Snelgrave, 125–7; Dalzel, xi, 126, 134, 176; Burton (1864; 1966 ed.), 111–12, 154–5, 166, 254–7, also the Terminal Essay to his translation of *The Arabian Nights*; Argyle, 87–8; Ajayi and Smith, 39.
21. Andrzejewski, chapter II.
22. Bowdich, 301, 315–6; McFarlane, 5. Bowdich's one-fifth would virtually equal the 21 per cent call-up of the French population in the First World War.
23. Andrzejewski, chapter VII.
24. Bosman, 397–8; Snelgrave, 56, 121.
25. Johnson, 451.
26. Law (1973, c), 2–3. For references to the distinction between heavy and light cavalry, and the propriety of making the distinction in West African history, see p. 95.
27. Denham and Clapperton (C), 47; Davies, 3.
27A. Barth, III, 165, 231, describes the Shuwa cavalry of mid-nineteenth-century Bornu as 'clad only in a loose shirt, and mounted upon their weak unseemly nags ... only armed with missiles usually consisting of one large spear, or kasákka, and and four small javelins, or bállem; very few of them have shields'.
28. See pp. 97, 112.
29. Davies, 3; Fisher (1972), 373; Fisher (1973), 362–4. Ibn Fartua (2)., 24, refers to the abandonment of the Bulala infantry by their cavalry after a defeat in battle.

30. Cited by Fage (1957), 82; Leo Africanus, 825.
31. Denham and Clapperton (D), 132, 160, 166–8 (and plate), 174; Barth, II, 329–30; III, 116 (plate), 149–50, 166. The plate in (D), engraved by Findon, are based on sketches by Denham himself.
32. Fisher and Rowland, 215–19.
33. Kea, 185–7, 207–8.
34. Tenkorang, *passim*.
35. Kea, 202.
36. Ajayi and Smith, 17–18.
37. R. Smith (1970), *passim* but especially 525–32.
38. Turney-High, 98.
39. Cited by Abdullahi Smith in Ajayi and Crowder, I, 200.
40. Tymowski, cited in R. Smith (1970), 530.
41. Barbot, 88, 154.
42. Landolphe, II, 98–100.
43. Bovill, 155; R. Smith (1970), 530.
44. Andrzejewski, chapter VII.
45. E. M. Chilver and P. M. Kaberry in Forde and Kaberry, 141–2; Meek (1937), 243.
46. Ibn Fartua, 30; Dapper, 237; M. G. Smith, 96; Meek (1931), 351; Burton (1864; 1966 ed.), 25, 360; Argyle, 84; Law (1971), 316–17; Hunwick in Ajayi and Crowder, I, 208; Echenberg, 242.
47. Adeleye in Ajayi and Crowder, I, 487;
47. Adeleye in Ajayi and Crowder, I, 487; M. G. Smith in Forde and Kaberry, 99; Zahan, ibid., 171–2; Law (1971), 319–25; Argyle, 88–9; Ryder, 8–9, 20; Tremearne, 151–83.
48. r. Smith (1970), 522, 528; Barth, III, 295.
49. De Graft Johnson and Datta, both *passim*.
50. Bradbury in Forde and Kaberry, 17, 28.
51. Bosman, 354.
52. See Goody, *passim*, especially chapter I.
52A. Law (1975), 13.
53. For examples, see M. G. Smith on Zazzau and Abuja, Nadel on Nupe, and Smaldone on the Sokoto Caliphate.
54. Bradbury in Forde and Kaberry, 8–12.
55. Allen and Thompson, I, 236.
56. Levtzion, 148; Ibn Fartua, 11–12. See also Brenner, 52.
57. Adeleye in Ajayi and Crowder, I, 506.
58. Law (1971), 320.
59. Zahan in Forde and Kaberry, 171–2.

60. Dalzel, x; Burton (1864; 1966 ed.), 264; Argyle, 86.
61. In Astley, III, 71.
62. Smaldone (1972), *passim*. See also Fisher and Rowland, *passim*.
63. Ajayi (1965), *passim*; R. Smith (1969), chapter XI; Ajayi and Smith, and Awe (1973), both *passim*.
64. Dalzel, 183; Law (1971), 333. Dalzel adds that two chiefs from Oyo and Allada respectively offered to provision the large camp set up by the Dahomeans at Badagry.
65. Bosman, 179–81. See also W. Smith, 217.
66. Barbot, 351; Norris, 137; Alagoa in Ajayi and Crowder, I, 299; Basden, 244, 382–4. Uka, 76–9, argues that the Abam warriors employed by the Aro were not mercenaries since their 'compelling desire' for war was to win heads in battle and the consequent heroic status of *ufiem*. But from the point of view of the Aro, who paid them, the Abam were their mercenaries or at least their agents.
67. Smith (1970), 527–8, 530–1.
68. J. R. Willis in Ajayi and Crowder, I, 451; Barth, III, 7–8, 61.
69. Dan Fodio, ii, 394.
70. Dupuis, xxxviii; H. Meredith in Wolfson, 94.
71. Barbot, 369.
72. Dapper, 301.
73. Captain Jones, cited and printed in Ajayi and Smith, 17, 134.
74. Denham and Clapperton (D), 165–6 and plate opposite 166; (C) 46; Muffett in Crowder (1971), 276.
75. Ibn Fartua (2), 40, 44; Barth, II, 155; III, 171, 175; Denham and Clapperton (D), 129; (C), 46; Law (1971), 346.
76. Burton (1864; 1966 ed.), 262–3; Burton (1863), I, 289; Snelgrave, 77.
77. Talbot (1923), 223.
78. Muffett in Crowder (1971), 280.
79. Johnson, 132; Burton (1864; 1966 ed.), 262; Bowdich, 301; Dupuis, 213; Denham and Clapperton (D), 129.
80. Dickenson, 120. See also Hopkins (1973), 30.
81. Barth, III, 184–5.
82. For examples from the Gold Coast, see Daaku, 175–6.

Chapter V

1. Turney-High, 5.
1A. Uka, 80, refers to Abam warriors blowing a powdered con-

coction known as *nkuku* on an enemy to make him sleepy, but does not say how this was done.

2. Goody, 43, 46.
3. De Marees in Wolfson, 53; Bosman, 128, echoed by Barbot, 262.
4. A chant by Baba Nino of Ile Jaati in Iwo, cited by Babalola (1966).
5. Ibn Fartua, 27.
6. Labat (1730), II, 242–3; Daaku, 152.
7. Goody, 28, 52.
8. Hopkins (1973), 81, fn. 7; Barth, II, 138–9.
9. Goody, 47–56.
10. Goody, 47.
10A. Kirk-Greene (1963), 174.
11. Barth, II, 45; Bivar, 13–27. Unfortunately Bivar omits to give the weight of these swords.
12. Ajayi and Smith, 134.
13. R. Smith (1971, b), 94.
14. It is at present in the writer's collection and was acquired from an Ibadan trader; it is illustrated in R. Smith (1967), 92.
15. For example, see R. Smith (1967), 91–4 and fig. 2.
16. Dapper, 185.
17. Goody, 35.
18. This point derives from Law (1973 b), 17–18. See also Law (1975), 6–7.
19. Barth, II, 45; Smaldone (1970), 24; Meek (1931), 447; Denham and Clapperton (D), plates 42, 43; Kirk-Greene (1963), *passim*; Muffett in Crowder (1971), 279, also 296, n. 15, which gives six Hausa words for the generic 'spear' and six for specialized types.
20. Denham and Clapperton (D), plate opposite 133 and plate 42; Jobson, 56. A Benin figure of a horseman which was sold at Sothebys in 1972 showed a lance in the right hand and a bundle of darts in the left (*Country Life*, 17 August 1972).
21. Ajayi and Smith, 134.
22. Denham and Clapperton (D), 132; Barth, III, 149–50.
22A. In Muffett (1964), 161. Several of the charging horsemen carry their spears above their heads, others are holding them horizontally.
23. Night watchmen in Northern Nigeria are still armed with bows and arrows and according to the Nigerian press bows

and arrows were used by the residents of Mushin near Lagos in January 1966 to repel political thugs.

24. Snelgrave, 56; Robertson, 206; Barth, II, 231–2; Smith (1967), 96 and fn. 28. Mounted bowmen were known in Europe and the Near East: for example, among the light cavalry of the Crusader Kingdom and, in the sixteenth century, the Yeomen of the Guard.

25. For example, Bosman, 186; Jobson, 57; Dapper, 236, 240, 301, 310.

26. Lander, II, 222; Muffett in Crowder (1971), 277; Echenberg, 248.

26A. In Muffett (1964), 128, there is a photograph of Arewa infantry presumably in the army of either Sokoto or Kano, holding bows, some of which appear to be as tall as the bowmen. But in Crowder (1971), 277, Muffett confirms the maximum height of the Caliphate bows as 'some five feet'.

27. *Rothmania maculata, Carpolobia latea, Cambretacea*, bamboo, are among the types: Smith (1967), 96; Muffett in Crowder (1971), 277; Meek (1931), 442.

28. Ellis, 172; Clapperton (1829), 118–19; Echenberg, 248–9, 253.

29. Ajayi and Smith, 84. See also Johnson, 350.

30. Moloney, *passim*; Balfour, *passim*; Talbot (1926), III, 825, 833; R. Smith (1967), 97. In the storerooms of the Nigeria Museum, Lagos, see G.46.16.29, 46.20.1, and 60.1.30A. As Richard Jefferies observes in *Bevis* (chapter XI), 'A crossbow..., to act well, must be made almost as accurately as a rifle'.

31. Illustrated in W. Fagg, plate 42.

32. Muffett in Crowder (1971), 296, n. 17, gives other ingredients of poisons with their Hausa names. An account of the poison used in Borgu is in Clapperton (1829), 83. See also La Chard, *passim*, and Echenberg, 249. Muffett (1964), 199, cites the case of a British officer who in the attack on Burmi in 1903 was struck above the knee by a poisoned arrow and died in twenty minutes.

33. Hiskett, 91–2; Clapperton (1829), 62; Smaldone (1970), 27; M. Smith (1960), 97.

34. Dapper, 311; Law (1973 b), 24; Law (1973 c), 12, n. 51.

35. A pair of these was presented by the writer to the museum of the University of Lagos in June 1973.

36. Tremearne, 184, notes the use of slings in war by the Kagoro. The difference between a sling and a catapult is

that the latter is used in conjunction with a forked stick.

37. Bosman, 396 (and Barbot, 336). Much the same account is given by Labat (1730), II, 244.

38. Cited in Ojo, 34.

39. White (1954), describes a number of these. See also Kirk-Greene (1972), 108, for a reference to 'boomarang knives' in Rabeh's army.

40. Talbot (1926), III, 823, and Basden, 387, both writing of Ibo warfare. Dapper, 313, describes how the Benin army defeated the cavalry of the 'Isago' to their westward by feigning retreat and drawing the enemy along paths where they fell into prepared pits covered with leaves. Law (1971), 339–40, identifies the Isago with the Oyo and suggests that the action took place somewhere in Ekiti country.

41. Ibn Fartua (2), 24; Davie, 3.

42. Denham and Clapperton (D), 64 (and plate), 129, 130, plate opposite 279; (C), 47; Barth, III, 165, 554; Smaldone (1970), 24, 32; Muffett in Crowder (1971), 278; Njeuma, 8. (The Bagirmi lancer in Denham's plate reminds the writer irresistibly of Tenniel's White Knight.)

43. Denham and Clapperton (D), 270.

44. Barth, II, 46 (with drawing), III, 129; Muffett in Crowder (1971), 279; Fisher (1973), 358.

45. Law (1973, c), 2, 9; Law (1975), 2, fn. 9, 6. But Law considers that 'a significant part of the Oyo cavalry fought without saddles and stirrups'.

46. Echenberg, 243; Talbot (1926), III, 833; R. Smith (1971, b), *passim*. Talbot writes that the armour of Oba Ozolua of Benin was so heavy that he could not lift it until he was about twenty years old, and that it was destroyed at the time of the British expedition to Benin in 1897.

47. Dapper, 301.

48. R. Smith (1967), 101.

49. Bivar, 12; Ibn Fartua, 15; A. Smith in Ajayi and Crowder, I, 195. For the location of Gaoga, see Kalck (1972), and subsequent comment in *JAH*, XIV (1973), 3.

50. Bivar, 10–12. The trade was two-way since hide for the manufacture of shields and probably hide shields themselves were exported from the Central Sudan to the north in the Middle Ages.

51. Ryder (1969), App. VIII; R. Smith (1971, b), 95–6.

52. Bosman, 186–7; Denham and Clapperton (D), 166, 168;

Barth, II, 45, 537, 540; Bivar, 9, 39–40; Muffett in Crowder (1971), 279; Fisher in Gray, 71. There are many surviving examples of shields from the savannah but only one from the forest has been traced by the writer, a hide shield preserved in the compound of the Shashire at Owo in Eastern Yorubaland.

53. R. Smith (1967), 100; Last, 52.
54. Dapper, 302; Labat (1730), II, 246; Ajayi and Smith, 42, 46, 86, 98; smaldone, 51. For drum signals, see Ibn Fartua (2), 38, 45.
55. Much of the following section is based on three articles in *JAH*, XII (1971), 2, by White, Kea and Fisher and Rowland.
56. Kea, 186, gives no grounds for his preference for the last reason. For the embargo on the provision of arms and ammunition to West Africa generally, see Mbaeyi (1974).
57. Ryder (1969), 68, 114. See Smith (1969), 124, for an uncorroborated tradition among the Ekiti that a Benin army attacked Ado, probably in the sixteenth century, with guns. This may reflect purchases from the Dutch.
58. Kea, 187–90, who adds that between the 1610s and 1650 the Dutch did not sell guns on the coast, and after 1650 sought to restrict their sale to those with whom they had military and commercial agreements. The English were prepared to sell to any trader.
59. Hopkins (1973), 110, 129. He considers that there was no causal relationship between the arms trade and the establishment of the slave trade.
60. Snelgrave, 56; Ajayi and Smith, 17–18. Law (1973, b), 19, points out that similarly 'the cavalry of Dagomba were able to fight at least a drawn campaign with the musketeers of Ashante' in 1744.
61. Davies, 4–5; Echenberg, 250.
62. William Smith, 182; Burton (1864; 1966 ed.), 53.
63. For example, see Snelgrave, 72.
64. Bosman, 184.
65. Ajayi and Smith, App. I.
66. Fisher and Rowland, 218, based on the *Kano Chronicle*.
67. Ibn Fartua, 11; Fisher in Gray, 70. Another possible explanation is contained in the relation by an anonymous Spaniard of the Moroccan expedition of 1590 to Songhay. The writer had heard how a Turkish army advancing at some previous time via Egypt had been defeated by the Bornu,

some 500 captive hand-gunners being subsequently enrolled in the Bornu army (information from Mr John Lavers).

68. ibid., 12, 27.
69. Fisher and Rowland, 217; personal communication from Mr John Lavers, based on the *Histoire chronologique du Royaume de Tripoly*.
70. Hiskett 78–9; Adeleye in Ajayi and Crowder, I, 504, 506; Barth, II, 45–6, 139.
71. Smaldone (1970), 89–96; (1972), 152–4.
71A. Johnson, 357, 415. The Prussian army adopted Dreyse's needlegun in 1848, while the French introduced the chassepot soon afterwards. Not until 1867 did the British military decide to convert their muzzle-loading Enfield rifles to breech-loaders on the Snider principle, and the Martini-Henry was adopted in 1871.
72. r. Smith in Crowder (1971), 181; Leggasick, 100–2; Person (1968), II, 907–10; Kirk-Greene (1972), 108. By this time a large proportion of Samori's rifles were repeaters, i.e. had magazines; most of then had been procured through Freetown.
72A. R. Smith (1967), 101; Clapperton (1829), 197.
73. Private communication from Professor Robin Horton; Johnson, 490. The allegation about the Ijesha was made in a letter from the Ibadan war leaders to prominent Oyo residents in Lagos.
74. Landolphe, I, 136. The passage describes the contrivance thus: 'un jeu d'orgues en cuivre armé de sept petits canons, en forme d'espingolle'.
75. The *Church Missionary Gleaner*, 1852, 115, has a drawing of a Dahomean cartouche belt.
76. Bosman, 187.
77. See, for example, Barbot, 397.
78. R. Smith (1970), 526–7.
79. Denham and Clapperton (D), 201, 208, 210, 249.
80. Saint-Martin, 794, 797; Kanya-Forstner in Crowder (1971), 57; Kirk-Greene (1972), 108.
81. Osifekunde in Curtin (1967), 287; Ajayi and Smith, 19–20, 134.
82. Dalzel, 10; Norris, 92; Burton (1864; 1966 ed.), 110, 173, 176, 181.
83. Public Record Office, F084/886, Beecroft to Palmerston/ Granville, 3 January 1852.

84. Barth, III, 212–13, 223. The Minié rifle with its expanding bullet had been introduced in 1851.

85. Ajayi and Smith, 139; Dupuis, 256 fn.; Burton (1864; 1966 ed.), 164; Legassick, 111; Smith in Crowder (1971), 181–2. Dupuis contrasts the Ashanti gun drill with that of the Fante who held their muskets away from their bodies and neglected to take aim.

85A. Burgess, 13–17. A 'major exception' to the pattern was the Congo Arabs.

86. Kea, 206; Afigbo, 81, for guns made by Awka smiths. Mr John Lavers has also drawn attention in a personal communication to a German report of the late eighteenth or early nineteenth century that 'metal cannon' were cast in Bornu by French slaves held by the Mai. The cannon were used by the Mai's army in wars against pagans to the south. (Seetzen material in von Zach, *Monatliche Correspondenz*, XXXIX (1810), 336.)

87. Kyerematen, 39.

88. Landolphe, I, 86; Hocquart in Hargreaves (1969), 191.

89. Legassick, 104; Person (1968), ii, 911–23. The repeaters were admired in 1891 by Archinard, the French commander, who commented on the smooth working of the breech mechanism.

90. Burton (1864; 1966 ed.), 102; Barth, III, 223. For further references, see Ajayi and Smith, 19, 40 fn., 134.

91. Echenberg, 252.

92. Denham and Clapperton (D), 75, 174.

93. Smaldone (1970), 108; Akintoye, 119 and fn. 45.

94. Crowder (1971), 8–9 and *passim*. Ukpabi (1970), 381, lists 'poor equipment, bad training (if they had any at all) and poor leadership' as responsible for the feeble resistance of West Africans to colonial troops.

95. See, for example, Bosman, 231–2; Dalzel, 21–2; Tremearne, 183, 186; Parkinson, 319–20.

96. Snelgrave, 78; Akinjogbin in Ajayi and Crowder, I, 324. The Ibadan cadets were known as *baba ni ng joko*, 'father says I should sit down', that is, on the battle-field. Wrestling matches like those described by Denham in Denham and Clapperton (D), 194–5, helped to keep the Bornu soldiers fit and alert.

97. Labat (1730), III, 233–4; Robertson, 270; Burton (1864; 1966 ed.), 63, 165–6, 204; Barth, II, 45.

98. Public Record Office, F084/1141, Foote to Russell, 9 February 1861, private.

99. Davies, 6–8.

99A. Rabeh's riflemen in Bornu were reported in 1894 to drill twice daily, while in 1901 the Emir of Kano was said to have sent slaves secretly to receive training in 'the arts of European warfare': Kirk-Greene (1972), 108, and Kirk-Greene and Newman (1971), 141.

100. Fisher (1972), 368–72. The earliest evidence, according to Law (1973, b), 12, is either the representation of what may be a horse (but is perhaps a donkey) on a fly-whisk handle excavated at Igbo Ukwu in East Central Nigeria, dated to about the ninth century A.D., or al-Muhallabi's mention of horses at Gao in the late tenth century. See also Afigbo, 81–2.

101. Fisher (1972), 377.

102. Law (1973, c), 2–3.

103. Leo Africanus, 825, 833. See also Ibn Fartua (2), 65, for Idris Aloma's purchases.

104. Fernandes in Hargreaves (1964), 16, 18; Law (1975), 3–4.

105. Clapperton (1829), 73; M. Smith, 137; Smaldone (1970), 27; Fisher (1972), 380, 385–6; J. Lavers, private communication. Barth, II, 315–6, praises the horses of Bornu. He implies that before 1854 (he does not say for how long) there had been an embargo on their export. In the horse fair at Kukawa, strong travelling horses for servants cost from six to eight dollars each, 'while an excellent horse would not fetch more than thirty dollars'. A good foreign horse cost 300 dollars.

106. Clapperton (1829), 73; Law (1975), 5.

107. Ojo, 112.

107A. In Sierra Leone, the spread of the virulent tsetse-borne *Trypanosoma brucei* has been associated with the deforestation which took place around Freetown in the nineteenth century: see Dorward and Payne (1975), *passim*. *G. palpalis* breeds commonly neat the oil palm while *G. longipalpis* prefers the secondary bush of the transition forest.

108. Fisher (1973), 386–7; Legassick, 106, citing Binger. Mr David Adeniji of Iwo kindly provided the writer with a list of medicines used for horses by the Yoruba. See also Law (1975), 5–6.

109. Law (1973, b), 21–2. Similarly, a car-owner today who does

not employ a chauffeur is derided in Lagos as an 'I-drive-myself'. See also Law (1975), 5.

110. Barth, III, 198; Fisher (1972), 381–2; Fisher (1973), 357. De Barros describes a display of horsemanship by some Wolof chiefs who had been taken to Portugal in the late fifteenth century: Hargreaves, 24.

111. Ibn Fartua, 33; Ibn Fartua (2), 55; Lavers, private communication.

112. Smaldone (1970), 51–2.

113. Last, 31, fn. 48.

114. Much of this section is based on R. Smith (1970), *passim*.

115. Alagoa (1970), 323–5.

116. Printed in Hunwick (1971), 580, who argues that it was unlikely that Kebbi was tributary to Songhay at this time.

117. Ibn Fartua, 33; Labat (1728), III, 83; Denham and Clapperton (D), 232; Barth, II, 324–6 and plate 331; IV, 374.

118. Sykes, 170, 178–81. Denham, in Denham and Clapperton (D), 60, also saw small plank-built boats on the river Yeou. According to Hopkins (1973), 179, the Buduma were able to dominate Lake Chad by their use of wooden canoes from 'the distant forest region to the south', for which they sold slaves, and they only took to reed canoes after colonial rule had ended the slave trade.

119. Rodney, 78 – though this in fact describes the purchase of boats by the *lançados*, or Portuguese settlers, on the Upper Guinea coast in the sixteenth to seventeenth centuries.

120. Barbot, 149, 152.

Chapter VI

1. Lloyd (1959), *passim*; Smith (1969), 79. The wall of Nanking, sometimes said to be the longest city wall in existence, has a circuit of some thirty miles.

2. Ibn Fartua, 30, 73; Siddle, *passim*; Smaldone (1970), 78; Johnson, 450; Akintoye, 12 and fn. 42; Uka, 80.

3. For a discussion of such walls in Yorubaland, see Ajayi and Smith, 23–6.

4. Robertson, 31.

5. Barth, IV, 123; Ajayi and Smith, 24; Akintoye, 43; Jeffreys, 275.

6. Cited in Palmer (1936), 229–30, fn. 3.
7. Tremearne, 138.
8. For examples, see Ajayi and Smith, 27–8 and 152–3 (plans 1 and 2); Barth, IV, 112 (and plan).
9. For examples, see Labat (1730), II, 306–7; Denham and Clapperton (D), 57; (C), 27; Barth, II, 37.
10. Denham and Clapperton (D), 230; (C), 50; Adeleye in Ajayi and Crowder,I, 515; Muffett (1964), 91. The town gate of Bauchi is now in the Nigeria Museum, Lagos. It is wholly of iron, hammered into narrow strips which have been riveted horizontally and vertically. It is said to have been made in Bauchi during the reign of Emir Yakubu I who died in 1845.
11. Jobson, 56; Denham and Clapperton (D), 230; Ajayi and Smith, 28, 137. See also Johnson, 209.
12. Denham and Clapperton (D), 230 (the double-walled citadel of Willighi); Clapperton (1829), 165 (the fortified palaces of Hausa rulers, with their battlements and arrow-slits).
13. Connah (1967), 608.
14. There is a tradition that some of the walls at Old Oyo were strengthened by the addition of palm wine to the earth.
15. See Ajayi and Smith, 26, fn. 3, for discussion of this.
16. Clapperton (1829), 98.
17. Clapperton (1839), 165–6, describes the house-building method of the Hausa. See also Moody, 27–8, for information about the use of the pear-shaped *tubali* (bricks) in the walls at Kano.
18. Connah (1967), 594.
19. Ajayi and Smith, Part I, chapter 3; Connah (1967), *passim*. See also Jobson, 55–6, for an example of a double palisade 'to keep off the force of horse'. An exception is the massive Eredo (p. 128 above), far longer than the wall of Ijebu Ode town.
20. Ozanne (1969), 29, found no evidence of a ditch outside the medieval walls of Ile Ife. This is exceptional.
21. Leslie, section 15, describing Ibikunle's addition to the walls of Ibadan.
22. J. D. Omer-Cooper, 'A preliminary report on the history of Owo', paper presented to the Congress of the Historical Society of Nigeria in 1960. But this suggestion is not repeated in Mabogunje and Omer-Cooper (1971).
23. Leo Africanus, 826; Denham and Clapperton (C), 135; Bowdich, 301; Norris, 13–14; Burton (1864; 1966 ed.), 171–2 (and

woodcut). It is recalled in the Egbado town of Ilaro that there were no walls; a nearby ravine was the traditional place of refuge for the inhabitants in time of war.

24. A. Smith (1970), *passim* and in Ajayi and Crowder, I, 187–91.
25. A. Smith in Ajayi and Crowder, I, 170, 182, 188 (fn. 105), 190.
26. Moody, 26.
27. Last, 36–7, 80.
28. Moody, 23–4. See also Fisher in Gray, 74–5.
29. Barth, II, 42, 52, 65, 118–20. See also Denham and Clapperton (C), 50, in which Clapperton gives the number of gates at Kano as fifteen. For photographs of the walls of Kano in 1903, see Muffett (1964), 80, 96–7.
30. Ibn Fartua, 16, 45. Ibn Fartua (2), 49, gives four reasons for the making of stockades (around camps): it enabled the animals to be tethered, frustrated thieves, prevented anyone leaving the camp for debauchery, and last (but surely not least) warded off a surprise attack by an enemy.
31. Last, 74.
32. Clapperton (1829), 181.
33. Ibid., 127. See also Barth, III, 121, 141, for a description of a Bornu camp. The quarters were 'light sheds constructed with the long stalks of Indian corn, supported by four poles ... forming high topped gables'.
34. Ozanne (1969), 28–37.
35. On Yoruba walls generally, see Ajayi and Smith, Part I, chapter 3; for the walls of Old Oyo, see Willett (1960, a); Ife, Willett (1960, b) and Ozanne (1969); Owo, Mabogunje and Omer-Cooper, 37–40. For the Benin walls, see Connah (1967) and his map (n.d.).
36. Connah (1967), 608.
37. Snelgrave, 28. Dalzel, 183, claims, hardly credibly, that the Dahomean camp at Badagry in 1784 was so large that it took four hours to traverse it on foot. See also Burton (1864; 1966 ed.), 333.
38. Ajayi and Smith, 28, 136.
39. The word 'Kiriji' is applied to the war and also to the camps and battlefields between the town of Igbajo and Imesi Ile. It was apparently coined to suggest the sound of the newly imported rifles when fired. The Yoruba are fond of these onomatopoeic derivations; cf. the Agidingbi, or 'booming', war at Lagos in 1851, imitating the sound of cannon, and the Pepe affair at Oyo in 1895, imitating the rattle of Bower's maxims.

40. Akintoye, 120, 132–7, and the contemporary photograph opposite 171.
41. Gleave (1963); R. Smith (1964); Law (1972); Clarke (ed. Atanda), chapter 3.
42. Turney-High, 17.
43. Denham and Clapperton (D), 235 and plate.
44. Ajayi and Smith, 148–9.
45. Roberton, 283.
46. Denham and Clapperton (D), 230.
47. Fynn, 71; R. Smith in Crowder (1971), 184–90.
48. Snelgrave, 10, 13.
49. Denham and Clapperton (D), 74, 168, 230.
50. Barth, III, 144; IV, 14.
51. Ahayi and Smith, Part I, chapter 7; Part II, chapters 4–6.
52. Dalzel, 201–2.
53. Ibid., 19; Burton (1864; 1966 ed.), 360.
54. Ibn Fartua, 27; Little in Forde and Kaberry, 243.
55. Cited by Hunwick in Ajayi and Crowder, I, 217.
56. Ibn Fartua, 24–8, cited in Palmer (1936), 241–2.
57. Hunwick in Ajayi and Crowder, I, 208.
58. Clapperton (1829), 185–90. 'Coonia' is possibly to be identified with Birni N'Konni, some fifty miles north of Sokoto, now in the Republic of Niger.
59. Ajayi and Smith, Part I, chapter 6.
60. Burton (1864; 1966 ed.), 359–64, based on the *Iwe Irohin* (the newspaper published by the missionaries at Abeokuta), April 1864. Burton had himself been in Abomey almost up to the time the expedition left for Abeokuta.
61. Ajayi and Smith, Part I, chapter 7; Part II, chapters 4–6; App. I (Captain Jones's report), and map 6.
62. Dalzel, 183–5.
63. Andrzejewski, 75–8. He suggests a connexion between the defensive tactics of the First World War with its trenches and fixed line machine guns and the subsequent establishment of small independent states.
64. Since this book went to press, the writer learnt from Mr Patrick Darling of the Benin Empire Iya Survey of a remarkable mosaic of walls (*iya*) surrounding existing or former towns and villages which extends at least 60 miles to the north-east of Benin City. These walls vary in height from about five feet to about 40 feet, with an average of about 12 feet. Some appear to be defensive in intention, but others seem equally clearly not

to be so, especially in cases where the walls of separate villages approach each other to within a few feet, leaving a narrow corridor between them.

Chapter VII

1. Clausewitz (ed. Rapoport), 402; see also 119.
2. Ibn Fartua, 5, 11, 30–1, 55.
3. Ibn Fartua 2, in Palmer (1967), 68–9; Fisher in Gray, 75.
4. Fynn, 74, 85–7, and thereafter *passim*; Boahen, 5; Boahen in Ajayi and Crowder, II, 170.
5. Dupuis, 257.
6. Atkins, 119–20; Dalzel, 7–8. See also Snelgrave, 5–6.
7. Akinjogbin (1963), 554; (1967), 73–7.
8. Law (1969), *passim*.
9. Morton-Williams (1964), *passim* and in Forde and Kaberry (1967), 39–40. See also R. Smith (1969), 45–6.
10. Cf. Morton-Williams (1964), ascribing a late eighteenth century date to the foundation of Ilaro and other Egbado towns, with Folayan (1967), who argues that they were established in the late sixteenth or early seventeenth century.
11. S. Johnson, 282; Ajayi and Smith, 29.
12. Smaldone (1970), 11–14.
13. Snelgrave, 9; Argyle, 83; Akinjogbin (1967), 63.
14. Bowdich, 301; Dupuis, 213. The Yoruba, Kagoro and, indeed, probably all armies wherever trees grow have made use of them for observation.
15. Barth, IV, 100–1.
16. See, for example, Ibn Fartua (2), 47, describing as 'spies' what was in fact a mounted Bulala patrol; also Meek (1931), 352, on Jukun reconnaissance.
17. Ajayi and Smith, 30, 137; Denham and Clapperton (D), 26, 43.
18. Ibn Fartua, 47; Tremearne, 183; Ricketts in Wolfson, 111; Nadel, 111.
19. Burton (1864; 1966 ed.), 288–9; Lombard in Forde and Kaberry, 88. Argyle maintains (84) that the Dahomeans put in their attacks just before dawn, but this is not borne out by their two attacks on Abeokuta.
20. Meek (1931), 351–2; Uka, 80.
21. Argyle, 88–9; Ajayi and Smith, 14–15.
22. Kea, 211.

23. Ibn Fartua (2), 35–6; Meek (1931), 352.
24. M. G. Smith (1960), 96.
25. Smaldone (1970), 116.
26. Denham and Clapperton (D), 162.
27. Akinjogbin (1967), 70; Burton (1864; 1966 ed.), 289.
28. Denham and Clapperton (D), 262.
29. For examples, see Nadel, 111; Smaldone (1970), 6, 55.
30. Labat (1730), 366; Denham and Clapperton (C), 47. See also Smaldone (1970), 55–6.
31. Denham and Clapperton (D), 174.
32. Barth, III, 149–50.
33. Dapper, 302; Barbot, 295–6; dan Fodio (ed. El-Masri), 492. For the gunfire and its effects, see Denham and Clapperton (D). 133 – 'the terror (the) miserable guns (of the Arabs) excited' – and Snelgrave, 56.
34. Ibn Fartua, 42; Bosman, 182; Labat (1730), I, 366–7. Writing of Samori's predominantly infantry army which fought the French at the end of the nineteenth century, Legassick, 110, uses language reminiscent of Labat's account: 'Major battles were fought by carefully arranged fixed lines. Once these were broken there was little hope of re-forming the army for further offence or defence on the same day.'
35. A skilled English archer in the fourteenth century, after shooting his first arrow, could follow it with five others before the first reached its target.
36. Ajayi and Smith, 30–1, 138–9.
37. Turney-High, 29–38.
38. Dan Fodio (ed. El-Masri), 488.
39. Echenberg, 243, writes of the Mossi in the late nineteenth century that they 'did not employ their cavalry as shock troops; horses and nobles were not so easily expendable as that.'
40. R. Smith in Crowder (ed.), (1971), 186–7.Muffett in Crowder (1971), 288–9, discusses the difficulties attendant on arming a cavalryman with a rifle. But there is no evidence of any use by mounted warriors in West Africa of firearms other than in their possible carrying of pistols.
41. Ibn Fartua, 16–21, 23. Idris Aloma's men concentrated on cutting down the enemy's trees in the dry season and his corn during the rains.
42. The following paragraphs are based on: Johnston, 44–6; Last, 26–7; Smaldone (1970), 6–8; Lafene and Boyd, *passim*.

The two contemporary accounts on which all these draw are the *Tazyin al-Waraqat* of Abdullahi Ibn Muhammad, translated and edited by M. Hiskett (see 112–13 for the battle), and the *Infaku'l Maisuri* of Muhammad Bello, translated by E. J. Arnett (1922). No account from the Gobir side is known.

43. *Infaku'l Maisuri*, cited by Johnston, 45–6.
44. Ibid., cited by Johnston, 46.
45. Last, 73.
46. Denham and Clapperton (D), 112, 133.
47. Ibid., 248–50.
48. S. Johnson, 222. Other examples might be taken from the attempts by the Nupe in the nineteenth century (when a part of the Fulani empire) to extend their rule over the north-eastern Yoruba, though here it was mainly the hilly nature of the country and the poisoned arrows of their adversaries which defeated them. A Benin plaque in the British Museum, probably dating from the early seventeenth century, shows infantry successfully resisting cavalry.
48A. Uka, 80.
48B. Ijagbemi, 250–1.
48C. Smith in Crowder (1971), *passim*.
49. Dupuis, 6–9.
50. Tenkorang, 3, 16.
51. Fynn, 143.
52. Clausewitz (ed. Rapoport), 117 (Book I, chapter I, section 21).
53. Cf. the outbreak of disease in the Axis army at El Alamein in 1942, which affected General Rommel himself and which has been blamed on the slackness in hygiene of the Italians.
54. Except where noted, the following paragraphs are based on R. Smith (1970), 525–32, and Ajayi and Smith, App. II.
55. Ibn Fartua, 45–6. He refers to these enemies of Mai Idris Aloma as the 'Talata' and the 'Kotoko'. These may be identified with the Yedina, called by the Kanuri 'Buduma', 'people of the reeds'.
56. Landolphe, I, 136.
57. Snelgrave, 9–19; Dalzel, 5, 14–19.
58. Norris, 54–6.
59. Dalzel, 168–9.
60. Rev H. Townsend, Journal (CMS, CA 085b), 17 March 1845.
61. See Lawrence (1963) and Van Dantzig and Priddy (1971), both *passim*.
62. Bosman, 14–18, 27–8.

63. Cited in Wolfson, 74.
64. Daaku, 76–7, 83, 130, App. II (deed of cession of 1624).
65. H. Meredith in Wolfson, 92–7.
66. An eye-witness account by Major H. J. Ricketts, *Narrative of the Ashantee War* (London, 1831), is reprinted in Wolfson, 110–15.
67. See the description by Reindorf reprinted in Wolfson, 115–19.
68. Akinjogbin (1967), 108.
69. W. Smith, 181–2.
70. Ajayi and Smith, 38; Kirk-Greene, 38.
71. Crowder (ed.) (1971), 16.

Chapter VIII

1. Goody, 47, 48, 51. A more convincing example of the 'founding horsemen' legend, on the other hand, is that concerning the origin of the Mossi state.
2. Lloyd in Lewis (1968), 35.
3. Awe (1973), 73–6. Echenberg, 247, describes the craft specialization developed in the standing armies of the Marka and Zaberma in the Upper Volta region in the late nineteenth century.
4. Hoebel, 331.
5. In Ajayi and Smith, 139.

Bibliography

I. **Contemporary Works**

'Abdullah Ibn Muhammad (ed. Hiskett, M.) (1963). *Tazyin al-Waraqat*. Ibadan.

Allen, Captain W. and Thompson, T. R. H. (1848). *Narrative of the Expedition to the River Niger in 1841*. London.

Astley, J. (1746. *Voyages and Travels*. London.

Atkins, J. (1735; 1737 ed.). *A Voyage to Guinea, Brazil and the West Indies*. London.

Barbot, J. (1732). *A Description of the Coasts of North and South Guinea and of Ethiopia Inferior, vulgarly Angola*. London.

Barth, H. (1857–8). *Travels and Discoveries in North and Central Africa*. London.

Bosman, W. (1705). *A New and Accurate Description of the Coast of Guinea*. London.

Bowdich, T. E. (1819). *Mission from Cape Coast Castle to Ashantee*. London.

Bowen, T. J. (1857). *Adventures and Missionary Labours in Several Countries in the Interior of Africa from 1849 to 1856*. Charleston.

Burton, Sir R. F. (1863). *Abeokuta and the Camaroons Mountain*. London.

—— (1864; 1966 ed.). *A Mission to Gelele King of Dahome*. London.

Clapperton, H. (1829). *Journal of a Second Expedition into the Interior of Africa from the Bight of Benin to Socattoo*. London.

Clarke, W. H. (ed. J. A. Atanda) (1972). *Travels and Explorations in Yorubaland 1854–1858*. Ibadan.

Dalzel, A. (1793). *The History of Dahomy, an Inland Kingdom of Africa*. London.

Dapper, O. (1686). *Description de l'Afrique*. Amsterdam.

Denham, D. and Clapperton, H. (1826). *Narrative of Travels and Discoveries in Northern and Central Africa in the Years 1822, 1823, and 1824*. London.

Duke, Antera. *Diary*. See Forde, D. (1956).

Dupuis, J. (1824). *Journal of a Residence in Ashantee*. London.

Forde, D. (1956). *Efik Traders of Old Calabar*. London.

Fodio, Usman dan. See El Masri, F. H. (1968) (unpublished).

Gibb, H. A. R. (1958; 1962). *The Travels of Ibn Battuta*. Cambridge.

Hiskett, M. (1963). See Abdullah Ibn Muhammad.

Houdas, O. See es Sa'di, 'Abd al-Rahman.

Ibn Battuta. See Gibb, H. A. R.

Ibn Fartua, translated by Palmer, Sir H. R. (1926). *History of the First Twelve Years of the Reign of Mai Idris Aloma of Bornu (1571–1583)*.

Ibn Fartua, A. (2). See Palmer, Sir H. R. (1967).

Ibn Khaldoun, A., translated de Slane, M. (1925). *Histoire des Berbères*. Paris.

Jobson, R. (1623). *The Golden Trade*. London.

Labat, J-B. (1728). *Nouvelle Relation de l'Afrique Occidental*. Paris.
——— (1730). *Voyage du Chevalier des Marchais en Guinée...* Paris.

Labarthe, P. (1803). *Voyage à la côte de Guinée*. Paris.

Lander, R. L. and J. (1832). *Journal of an Expedition to Explore the Course and Termination of the Niger*. London.

Landolphe, Captain J. F. See Quesne, J. S.

Leo Africanus (1896), translated by Poris, J., and ed. Brown, R. *The History and Description of Africa*. London.

Norris, R. (1789). *Memoirs of the Reign of Bossa Ahadee King of Dahomy*. London.

Palmer, Sir H. R. (1928; 1967 reprint). *Sudanese Memoirs*. London (for Ibn Fartua (2)).

Quesne, J. S. (1823). *Mémoires du Capitaine Landolphe...* Paris.

Robertson, G. A. (1819). *Notes on Africa*. London.

Sa'di, A. es-. translated Houdas, O. (1900). *Tarikh es-Sudan*. Paris.

Smith, W. (1764). *A New Voyage to Guinea*. London.

Snelgrave, Captain W. (1734). *A New Account of Some Parts of Guinea and the Slave Trade*. London.

Tully, R. (1817, second ed.). *Narrative of a Ten Years Residence at Tripoli in Africa*. London.

II. Modern works

A. *Books*

Adandé, A. (1962). *Les Récades des Rois du Dahomey*. Dakar.

Adeleye, R. A. (1971). *Power and Diplomacy in Northern Nigeria 1804–1906*. London.

Ajayi, J. F. A. and Crowder, M. (1971; 1974) (ed.). *History of West Africa*, Volumes I and II. London.

―――― and Smith, R. (1964; 1971 ed.). *Yoruba Warfare in the Nineteenth Century*. Cambridge.

Ajisafe, A. K. (1924). *The Laws and Customs of the Yoruba People*. London.

―――― (1945). *Laws and Customs of the Benin People*. Lagos.

Akinjogbin, I. A. (1967). *Dahomey and its Neighbours, 1708–1818*. Cambridge.

Akintoye, S. A. (1971). *Revolution and Power in Yorubaland, 1840–1893*. London.

Allott, A. (1960). *Essays in African Law*. London.

Andrzejewski, S. (1954). *Military Organization and Society*. London.

Argyle, W. J. (1966). *The Fon of Dahomey*. Oxford.

Arnett, E. J. (1922). *The Rise of the Sokoto Fulani*. Kano.

Basden, G. T. (1938; 1966 ed.). *Niger Ibos*. London.

Biobaku, S. O. (ed.) (1973). *The Sources of Yoruba History*. Oxford.

Bivar, A. D. H. (1964). *Nigerian Panoply. Arms and Armour of the Northern Region*. Lagos.

Boahen, A. A. See Ingham, K.

Bohannan, L. and P. (1953; 1962 ed.). *The Tiv of Central Nigeria*. London.

Claridge, W. W. (1915; 1966 reprint). *A History of the Gold Coast and Ashanti*. London.

Clausewitz, C. von (1832; 1968 ed. Rapoport, A.). *On War*. London.

Crowder, M. (ed.) (1971). *West African Resistance*. London.

Curtin, P. D. (ed.) (1967). *Africa Remembered*. Wisconsin.

Daaku, K. Y. (1970). *Trade and Politics on the Gold Coast, 1600–1720*. Oxford.

Dalziel, J. M. (1937). *The Useful Plants of Tropical West Africa*. London.

Dantzig, A. van and Priddy, B. (1971). *A Short History of the Forts and Castles of Ghana*. Accra.

Davie, M. R. (1929). *The Evolution of War*. Yale.

Diamond, A. S. (1935; 1950 ed.). *Primitive Law*. London.

——— (1971). *Primitive Law, Past and Present*. London.

Dickson, K. B. (1969). *A Historical Geography of Ghana*. Cambridge.

Elias, T. O. (1956). *The Nature of African Customary Law*. Manchester.

Ellis, A. B. (1894). *The Yoruba Speaking Peoples of the Slave Coast of West Africa*. London.

Fagg, W. (1963). *Nigerian Images*. New York.

Falls, C. (1961). *The Art of War*. London.

Forde, D. and Kaberry, P. M. (ed.) (1967). *West African Kingdoms in the Nineteenth Century*. Oxford.

Fortes, M. and Evans-Pritchard, E. E. (1940). *African Political Systems*. Oxford.

Fraser, D. and Cole, H. M. (ed.) (1972). *African Art and Leadership*. Wisconsin.

Fynn, J. K. (1971). *Asante and its Neighbours*. London.

Gilbert, M. (ed.) (1966). *A Century of Conflict 1850–1950*. London.

Goody, J. (1971). *Technology, Tradition, and the State in Africa*. London.

Gray, R. (ed.) (1975). *The Cambridge History of Africa*, Vol. 4.

Hargreaves, J. D. (1969). *France and West Africa*. London.

Hayford, C. (1903; 1970 ed.). *Gold Coast Native Institutions*. London.

Hiskett, M. (1973). *The Sword of Truth*. London.

Hodgkin, T. See Ingham, K.

Hoebel, E. A. (1954). *The Law of Primitive Man*. Harvard.

Hopkins, A. G. (1973). *An Economic History of West Africa*. London.

Ingham, K. (ed.) (1974). *The Foreign Relations of African States*. London.

Johnson, Rev S. (1921). *The History of the Yorubas*. Lagos.

Johnson, H. A. S. (1967). *The Fulani Empire of Sokoto*. Oxford.

Khadduri, M. (1955). *War and Peace in the Law of Islam*. Baltimore.

Kirk-Greene, A. H. M. (1958). *Adamawa, Past and Present*. London.

——— (ed.) (reprint, 1972). *Gazeteer of the Northern Provinces of Nigeria*, Vol. II. London.

Kirk-Greene, A. H. M. and Newman, P. (eds.) (1971). *West African Travels and Adventures*. Yale.

Kyerematen, A. A. Y. (1964). *Panoply of Ghana*. London.

Larsen, A., Jenks, C. W. and others (1965). *Sovereignty within the Law*. New York.

Last, M. (1967). *The Sokoto Caliphate*. London.

Lawrence, A. W. (1963). *Fortified Trade Posts. The English in West Africa, 1645–1822*. London.

Levtzion, N. (1973). *Ancient Ghana and Mali*. London.

Lewis, I. M. (ed.) (1968). *History and Social Anthropology*. London.

Lloyd, P. C. (1971). *The Political Development of Yoruba Kingdoms in the Eighteenth and Nineteenth Centuries*. London.

McCall, D. F and Bennett, N. R. (1971). *Aspects of West African Islam*. Boston.

McCulloch, M. (1950). *Peoples of Sierra Leone*. London.

Mattingly, G. (1955; 1965 ed.). *Renaissance Diplomacy*. London.

Meek, C. K. (1931). *A Sudanese Kingdom*. London.

——— (1937). *Law and Authority in a Nigerian Tribe*. Oxford.

Muffett, D. J. M. *Concerning Brave Captains*. London.

Nadel, S. F. (1942). *A Black Byzantium*. Oxford.

Newbury, C. W. (1961). *The Western Slave Coast and its Rulers*. Oxford.

Nicolson, H. (1939). *Diplomacy*. Oxford.

Numelin, R. (1950). *The Beginnings of Diplomacy*. Oxford.

Nys, E. (1884). *Les Origines de la Diplomatie et le droit d'Ambassade jusqu'à Grotius*. Brussels.

Ojo, G. J. A. (1966). *Yoruba Culture*. London.

Palmer, Sir H. R. (1936). *The Bornu Sahara and Sudan*. London.

Person, Y. (1968). *Samori. une revolution Dyula*. Dakar.

Priddy, B. See van Dantzig.

Rapoport, A. See Clausewitz, C. von.

Rodney, W. (1970). *A History of the Upper Guinea Coast 1545–1800*. Oxford.

Ryder, A. F. C. (1969). *Benin and the Europeans, 1485–1897*. Oxford.

Smith, M. G. (1960). *Government in Zazzau 1800–1950*. Oxford.

Smith, R. See Ajayi and Smith.

Smith, R. (1969; 1976 ed.). *Kingdoms of the Yoruba*. London.

Sundström, L. (1965). *The Trade of Guinea*. Lund.

Sykes, S. K. (1972). *Lake Chad*. London.

Talbot, P. A. (1923). *Life in Southern Nigeria*. London.

——— (1926). *The Peoples of Southern Nigeria*. London.

Trimingham, J. S. (1959). *Islam in West Africa*. Oxford.
Turney-High, H. H. (1949; 1971 ed.). *Primitive War*. Carolina.
Verger, P. (1968). *Flux et Reflux de la Traite des Nègres* ... Paris.
Willett, F. (1971). *African Art. An Introduction*. London.
Wolfson, F. (1958; 1962 ed.). *Pageant of Ghana*. London.
Wright, Q. (1942; 1965 ed.). *A Study of War*. Chicago.

B. *Articles*

Afigbo, A. E. (1973). 'Trade and trade routes in nineteenth century
 Nsukka', *J. Hist. Soc. Nigeria*, VIII, 1.
Ajayi, J. F. A. (1965) 'Professional Warriors in Nineteenth Century
 Yoruba Politics', *Tarikh*, I, 1.
———— and Austen, R. A. (1972). 'Hopkins on Economic Im-
 perialism in West Africa', *Econ. Hist. Review*, XXV, 2.
Akinjogbin, I. A. (1963). 'Agaja and the Conquest of the Central
 Aja States, 1724–1730', *J. Hist. Soc. Nigeria*, II, 4.
Alagoa, E. J. (1970). 'Long-distance trade and States in the Niger
 Delta', *J. African Hist.*, XI, 3.
Arhin, K. (1967). 'The Structure of Greater Ashanti (1700–1824)',
 J. African His., VIII, 1.
Awe, B. (1973). 'Militarism and Economic Development in Nine-
 teenth Century Yoruba Country: the Ibadan Example',
 J. African Hist., XIV, 1.
Balfour, H. (1908–9). 'The Origin of West African Crossbows',
 J. African Soc.
Bivar, A. D. H. and Hiskett, M. (1962). 'The Arabic Literature of
 Nigeria to 1804: a provisional account', *Bull. S.O.A.S.*,
 XXV.
Bloxham, G. W. (1887). 'Exhibition of West African symbolic
 messages', *J. Anthrop. Inst.*, XVI.
Bontinck, F. (1970). 'La Premiere "Ambassade" Congolaise à
 Rome', *Études d'Histoire africaine*, I.
Brenner, L. (1973). 'Sources of constitutional thought in Bornu',
 J. Hist. Soc. Nigeria, VII, 1.
Brown, P. (1951). 'Patterns of Authority in West Africa', *Africa*,
 XXI, 4.
Collins, E. (1962). 'The Panic Element in Nineteenth Century
 British Relations with Ashanti', *Trans. Hist. Soc. Ghana*,
 V, 2.
Connah, G. (1967). 'New Light on the Benin City Walls', *J. Hist.
 Soc. Nigeria*, III, 4.

Curtin, P. D. (1971). 'Jihad in West Africa: early phases and inter-relations in Mauretania and Senegal', *J. African Hist.*, XII, 1.

Datta, A. (1972). 'The Fante Asafo: a re-assessment', *Africa*, XLII, 4.

Dorjahn, V. R. and Fyfe, C. (1962). 'Landlord and Stranger: Change in Tenancy Relations in Sierra Leone', *J. African Hist.*, III, 3.

Dorward, D. C. and Payne, A. I. (1975). 'Deforestation, the decline of the horse, and the spread of the tsetse fly and trypanosomiasis (*nagana*) in nineteenth century Sierra Leone', *J. African Hist.*, XVI, 2.

Echenberg, M. J. (1971). 'Late Nineteenth-Century Military Technology in Upper Volta', *J. African Hist.*, XII, 2.

Fage, J. D. (1957). 'Ancient Ghana. A Review of the Evidence', *Trans. Hist. Soc. Ghana*, III, 2.

——— (1967). 'On the reproduction and editing of classics of African history', *J. African Hist.*, VIII, 1.

Farrer, J. A. (1880). 'Savage and Civilised Warfare', *J. Anthrop. Inst.*, IX.

Fisher, H. J. (1971). 'Prayer and Military Activity in the History of Muslim Africa South of the Sahara', *J. African Hist.*, XII, 3.

——— (1972–1973). ' "He swalloweth the ground with fierceness and rage": the horse in the Central Sudan', I and II, *J. African Hist.*, XIII, 3, XIV, 3.

——— and Rowland, V. (1971). 'Firearms in the Central Sudan', *J. African Hist.*, XII, 2.

Folayan, K. (1967). 'Egbado to 1832: the Birth of a Dilemma', *J. Hist. Soc. Nigeria*, IV, 1.

Gleave, M. B. (1963). 'Hill Settlements and their Abandonment in Western Yorubaland', *Africa*, XXXIII, 4.

Hagan, G. P. (1971). 'Ashanti Bureaucracy', *Trans. Hist. Soc. Ghana*, XII.

Hallam, W. K. R. (1970). 'The Battle of Fashegar', *Nigeria Magazine*, 106.

Hopkins, A. (1968). 'Economic Imperialism in West Africa', *Econ. Hist. Review*, Second Series, XXI, 3.

——— (1972). 'Economic Imperialism in West Africa: a Rejoinder', *Econ. Hist. Review*, Second Series, XXV, 2.

Hunwick, J. O. (1971). 'A little-known diplomatic episode in the history of Kebbi (*c.* 1594)', *J. Hist. Soc. Nigeria*, V, 5.

Ijagbemi, A. (1974). 'The Yoni Expedition of 1887: a study of British Imperial expansion in Sierra Leone', *J. Hist. Soc. Nigeria*, VII, 2.

Jeffreys, M. D. W. (1961). 'Some historical notes on the Ntem', *J. Hist. Soc. Nigeria*, II, 2.

Johnson, J. C. de G. (1932). 'The Fante Asafu', *Africa*, V, 3.

Kalck, P. (1972). 'Pour une localisation du Royaume de Gaoga', *J. African History*, XIII, 4.

Kea, R. A. (1971). 'Firearms and Warfare on the Gold and Slave Coasts from the seventeenth to the nineteenth centuries', *J. African History*, XII, 2.

Kirk-Greene, A. H. M. (1959). 'The British Consulate at Lake Chad', *African Affairs*, 58.

——— (1963). 'A note on some spears from Bornu, Northern Nigeria', *Man*, 220.

——— (1964). 'A preliminary note on new sources for Nigerian military history', *J. Hist. Soc. Nigeria*, III, 1.

La Chard, L. W. (1905–6). 'The arrow poisons of Northern Nigeria', *J.R. African Inst.*, V.

Lafene, J. E. and Boyd, J. F. (1970). 'A Visit to Tabkin Kwatto', *Nigeria Magazine*, 106.

Law, R. C. C. (1969). 'The fall of Allada: an ideological revolution?', *J. Hist. Soc. Nigeria*, V, 1.

——— (1972). 'Iwere', *J. Hist. Soc. Nigeria*, VI, 2.

——— (1973 a). 'Anthropological models in Yoruba history', *Africa*, XLIII, 1.

——— (1975). 'A West African Cavalry State: the Kingdom of Oyo', *J. African History*, XVI, 1.

Legassick, M. (1966). 'Firearms, horses and Samorian army administration', *J. African History*, VII, 1.

Levtzion, N. (1965). 'Early nineteenth century Arabic manuscripts from Kumasi', *Trans. Hist. Soc. Ghana*, VIII.

Little, K. (1965–66). 'The political function of the Poro', *Africa*, XXXV, 4, XXXVI, 1.

Lloyd, P. C. (1959). 'Sungbo's Eredo', *Odu*, 7.

Lovejoy, P. E. (1971). 'Long-distance trade and Islam: the case of the nineteenth century Hausa kola trade', *J. Hist. Soc. Nigeria*, V, 4.

McFarlane, K. B. (1962). 'War, the economy and social change', *Past and Present*, 22 (reporting a conference on 'War and society, 1300–1600').

Markov, W. and Sebald, P. (1967). 'The treaty between Germany and the Sultan of Gwandu', *J. Hist. Soc. Nigeria*, IV, 1.

Martin, B. G. (1962). 'Five letters from the Tripoli archives', *J. Hist. Soc. Nigeria*, II, 3.

Mbaeyi, P. (1974). 'Arms and ammunition, and their embargo in British West African history, 1823–1874', *Ikenga*, II, 2.

Moloney, Sir A. (1890). 'On Cross-bow, Long-bows, Quivers, etc. from the Yoruba Country', *J. Anthrop. Inst.*, XIX.

Moody, H. B. L. (1967). 'Ganuwa – the Walls of Kano City', *Nigeria Magazine*, 92.

Morton-Williams, P. (1964). 'The Oyo Yoruba and the Atlantic Trade, 1670–1830', *J. Hist. Soc. Nigeria*, III, 1.

Njeuma, M. Z. (1973). 'The foundations of pre-European administration in Adamawa: historical considerations', *J. Hist. Soc. Nigeria*, VIII, 1.

Ozanne, P. (1969). 'A New Archaeological Survey of Ife', *Odu* (new series), 1.

Parkinson, J. (1906). 'Notes on the Asaba People (Ibos) of the Niger', *J. R. Anthrop. Inst.*, XXXVI.

Rainero, R. (1966). 'La Bataille de Tondibi (1591) et la conquête marocaine de l'Empire Songhay', *Genère-Afrique*, V, 2.

Rowland, see Fisher and Rowland.

Ryder, A. F. C. (1961). 'The Benin Missions', *J. Hist. Soc. Nigeria*, II, 2.

Saint-Martin, Y-J. (1968). 'La volonté de paix d'El Hadj Omar et d'Ahmadou dans leurs relations avec la France', *Bull. de l'IFAN*, XXX, B, 3.

Siddle, D. J. (1968). 'War-towns in Sierra Leone: a study in social change', *Africa*, XXXVIII, 1.

Smaldone, J. P. (1972). 'Firearms in the Central Sudan: a Revaluation', *J. African Hist.*, XIII, 4.

Smith, H. F. C. (Abdullahi) (1961). 'The Islamic Revolutions of the 19th Century', *J. Hist. Soc. Nigeria*, II, 2.

Smith, Abdullahi (1971). 'Some considerations relating to the formation of states in Hausaland', *J. Hist. Soc. Nigeria*, V, 3.

Smith, R. (1964). 'Erin and Iwawun, Forgotten Towns of the Oke Ogun', *Odu*, I, 1.

———— (1967). 'Yoruba Armament', *J. African Hist.*, VIII, 1.

———— (1969). 'To the Palaver Islands: war and diplomacy on the Lagos lagoon, 1852–1854', *J. Hist. Soc. Nigeria*, V, 1.

———— (1970). 'The Canoe in West African History', *J. African Hist.*, XI, 4.

———— (1971, a). 'Event and Portent: the Fall of Old Oyo, a Problem in Historical Explanation', *Africa*, XLI, 3.

———— (1971, b). 'A Note on Two Shirts of Chain Mail in the Palace at Owo in South-western Nigeria', *Odu*, (n.s.), 5.

———— (1973). 'Peace and Palaver: international relations in pre-colonial West Africa', *J. African Hist.*, XIV, 4.

Tenkorang, S. (1968). 'The Importance of Firearms in the Struggle between the Ashanti and the Coastal States, 1708–1807', *Trans. Hist. Soc. Ghana*, IX.

Tremearne, A. J. N. (1912). 'Notes on the Kagoro and other Niger Head-Hunters', *J. R. Anthrop. Inst.*, XLII.

Tymowski, M. (1967). 'Le Niger, voie de communication des grands états du Soudan occidentale jusqu'à la fin du XVIè. siècle', *Africana Bull.*, 6, Warsaw University.

Uka, N. (1972). 'A note on the "Abam" warriors of Igboland', *Ikenga*, I, 2.

Ukpabi, S. C. (1970). 'British colonial wars in West Africa: image and reality', *Civilisation*, XX, 3.

Verger, P. (1970). 'Échanges de cadeaux entre rois d'Abomey et souverains européens aux XVIIIe et XIXe siècles', *Bull. de l'IFAN*, XXII, B, 3.

White, S. (1954). 'A study of Northern Throwing Knives', *Nigeria Magazine*, 44.

White, Rev G. (1971). 'Firearms in Africa: an introduction', *J. African Hist.*, XII, 2.

Willett, F. (1960 a). 'Investigations at Old Oyo, 1956–57: an Interim Report', *J. Hist. Soc. Nigeria*, II, 1.

———— (1960, b). 'Ife and its Archaeology', *J. African Hist.*, I, 2.

C. *Unpublished Works*

Babalola, A. (1966). 'Rara Chats in Yoruba Spoken Art', Seminar paper, University of Lagos.

Burgess, A. S. (1970). 'The arms trade in East Africa, 1840–1890', unpublished essay, University of Aberdeen.

Davies, – (c. 1930?). 'The History and Organization of the "Kambonse" in Dagomba'.

Law, R. C. C. (1971). *The Oyo Empire: the history of a Yoruba state, principally in the period c. 1600–c. 1836*, Ph.D. thesis, University of Birmingham.

—— (1973, b). 'The Horse in Pre-Colonial West Africa', Seminar paper, University of Stirling.

—— (1973, c). 'Horses, Firearms, and Political Power in Pre-Colonial West Africa', Seminar paper, School of Oriental and African Studies, University of London.

El-Masri, F. H. (1968). *A Critical Edition of Dan Fodio's Buyan Wujub al-hijra 'ala .l-'Ibad*, Ph.D. thesis, University of Ibadan.

Omer-Cooper, J. D. (1960). 'A Preliminary Report on the History of Owu, an ancient Yoruba city', paper presented before the Congress of the Historical Society of Nigeria.

*Smaldone, J. P. (1970). *Historical and Sociological Aspects of Warfare in the Sokoto Caliphate*, Ph.D. thesis, North-western University.

*Since this book went to press, J. P. Smaldone's *Warfare in the Sokoto Caliphate: Historical and Sociological Perspectives* (Cambridge) has been announced.

Index